NCSM ESSENTIAL ACTIONS

Framework for Leadership in Mathematics Education

NCSM

LEADERSHIP IN MATHEMATICS EDUCATION

NCSM ESSENTIAL ACTIONS SERIES

NCSM

LEADERSHIP IN MATHEMATICS EDUCATION

The permission can be obtained by writing to:
NCSM, P.O. Box 3406, Englewood, CO 80155
or email office@mathedleadership.org

First paper-back edition March 2020
ISBN: 978-0-9890765-5-5

Book cover designed by Allegra www.allegradenverdtc.com
Book edited by Gillian Dickens
Original Graphics by James Tilley
Final Graphics and Book Layout by Lori Flores
Printed by Allegra Marketing • Print • Mail
Published by NCSM: Leadership in Mathematics Education

Credits and Acknowledgments

Framework of Leadership in Mathematics Education is the third book in our *NCSM Essential Actions Leadership Series.* In 2017, the NCSM board revised its mission and vision to reflect the core values and beliefs of its membership and the changing needs of our students, teachers, and leaders. Our bottom line is to provide "high-quality mathematics teaching and learning every day for each and every learner." This is not possible without collaboration and cooperation among all stakeholders.

At the annual conference in 2019, NCSM released the first two books in the series, *Coaching in Mathematics Education* and *Instructional Leadership in Mathematics Education.* The first book defines the vision of mathematics coaching and the foundational elements that should be present in a high-quality coaching program. The book also details the essential actions needed to maintain an effective mathematics coaching program. The second book addresses the essential actions that principals, as instructional leaders of mathematics, should follow in order to promote high-quality mathematics teaching and learning at the site level.

The third book in the series builds on the two guiding NCSM publications, *The PRIME Leadership Framework* and *It's TIME: Themes and Imperatives for Mathematics Education,* to define the essential actions for all mathematics leaders, regardless of their title.

Leaders in mathematics education have a variety of titles, roles, and responsibilities. However, our titles do not make us leaders; rather, our actions positively impact those we serve define us as leaders. The essential actions described in this book and throughout the rest of the series bring focus and clarity to both the actions mathematical leaders employ to support all learners and the strategies and tools to build the essential leadership actions.

On behalf of our entire board and membership, I would like to thank the writing team for their countless volunteer hours they dedicated to this NCSM endeavor. In addition, I want to personally thank the mathematics leaders who reviewed, supported, and cheered on this work and through those efforts helped shape this book.

We are honored to share this framework and provide support as you boldly lead mathematics.

Mona Toncheff
NCSM President 2019–2021

NCSM Writing Team

Paul Gray, Lead Writer
Bill Barnes
Brian Buckhalter
Linda Fulmore
Linda Griffith
Christina Lincoln-Moore
Shelly M. Jones
Michelle Rinehart
Sarah Schuhl
John Staley
Tom Stricklin
Mona Toncheff
Shawn Towle
Desha Williams
Gwen Zimmermann

Reviewers

Hilary Kreisberg
Steve Leinwand
April Pfouts
NCSM Board 2019-2020

Table of Contents

About NCSM

NCSM is a premier leadership organization for those that serve the NCSM mission and vision for equity and excellence for student achievement in mathematics. NCSM is founded on the strength and dedication of a growing membership of mathematics education leaders. These leaders include grade-level team leaders; course-level team leaders; department chairs; district or county coaches; site-based teacher leaders; district, state, or provincial leaders; college faculty; and all who work to ensure the success of each and every mathematics student.

NCSM was created at the 1968 Philadelphia meeting of the National Council of Teachers of Mathematics (NCTM), when a group of urban district supervisors decided at the next annual meeting (Minneapolis, 1969) that school district leaders should gather together to form what is now called NCSM to specifically address leadership issues in mathematics. The members started with mathematics supervisors, and an early critical issue for NCSM was defining the membership. The founding members chose *not* to restrict members to supervisors and instead welcomed all leaders and teachers of mathematics. This theme has continued throughout the years; the 35 leaders who attended the first meeting in Minneapolis grew to more than 3,000 by the end of the 20th century.

As NCSM celebrated its 50th anniversary in 2018, the mission and vision for the organization were revised to represent the core values and beliefs of its membership and the changing needs of our students, teachers, and leaders.

NCSM Mission Statement

NCSM is a mathematics education leadership organization that equips and empowers a diverse education community to engage in leadership that supports, sustains, and inspires high-quality mathematics teaching and learning every day for each and every learner.

NCSM Vision Statement

NCSM is the premiere mathematics education leadership organization. Our bold leadership in the mathematics education community develops vision, ensures support, and guarantees that all students engage in equitable, high-quality mathematical experiences that lead to powerful, flexible *uses of mathematical understanding to affect their lives and to improve the world.*

High-quality leadership is vital to this vision. NCSM is committed to:

Developing and Informing Vision

- Provide leadership to influence issues and policies affecting mathematics education in ways consistent with the mission and vision of NCSM.
- Equip leaders to be critical consumers of educational information, research, and policy to become change agents in their communities.
- Support leaders to develop an actionable vision of mathematics instruction consistent with a view of mathematics as a sense-making endeavor.

Ensuring Support to All Stakeholders

- Develop networking and communication opportunities that connect the mathematics education community, as well as the broader education community.
- Equip leaders with the tools to create and sustain systems that fully align with the vision of mathematics and mathematics instruction promoted by NCSM.
- Equip leaders with the understanding, knowledge, and skills to continue their own personal growth, support emerging leaders, and further develop excellence in mathematics teaching.

Guaranteeing All Students Engage in Equitable, High-Quality Mathematical Experiences

- Provide advocacy and support regarding issues and policies affecting mathematics education in ways consistent with the mission and vision of NCSM.
- Provide resources for implementation of research-informed instruction to ensure students engage in relevant and meaningful learning experiences that promote mathematics as a sense-making endeavor.
- Advocate for each and every student to have access to rigorous mathematics that develops their understanding, skills, and knowledge, along with the confidence to leverage their learning, in order to improve their world.

As NCSM enters its sixth decade, we are striving for bold leadership actions to *ensure* meaningful and relevant mathematics for each and every learner. Our greatest challenge is to strive for meeting our vision by developing the leadership knowledge that was set forth in *The PRIME Leadership Framework* (NCSM, 2008) and the Leadership Framework in *It's TIME: Themes and Imperatives for Mathematics Education* (NCSM, 2014b). How can we develop a shared vision of high-quality, meaningful, and relevant mathematics? How can we ensure all stakeholders understand the actions and have the knowledge and skills to deliver meaningful and relevant mathematics? Can we, the leaders of mathematics, guarantee student engagement in equitable, high-quality mathematics experiences? Yes, we can, and we will because we are NCSM.

—The NCSM Board (2019–2021)

Introduction

NCSM is the premiere mathematics education leadership organization. Our bold leadership in the mathematics education community develops vision, ensures support, and guarantees that all students engage in equitable, high-quality mathematical experiences that lead to powerful, flexible uses of mathematical understanding to affect their lives and to improve the world.

—NCSM Vision Statement, 2017

The *Framework for Leadership in Mathematics Education* describes four guiding principles, each with three specific imperatives, that, when employed together, lead to increased student achievement in mathematics. The NCSM vision clarifies the focus on student learning outcomes and directs the charge to "guarantee that all students engage in equitable, high quality mathematical experiences." This is a tall order. So, how do we bring the vision to life? How do we ensure equitable student engagement in mathematics classrooms of the teachers we serve? How do mathematics education leaders work together with all community stakeholders to bring this vision to life?

If our goal is to provide the best learning experiences for all students, then we have no choice but to keep equity at the forefront of our minds. Consider the wisdom of Rochelle Gutiérrez:

> Equity in mathematics education will not be achieved until it is no longer possible "to predict mathematics achievement and participation based solely on student characteristics such as race, class, ethnicity, sex, beliefs, and proficiency in the dominant language." (Gutiérrez, 2002, p. 153)

Educators have known for decades achievement gaps among various student populations persist. At NCSM's Annual Meeting in 2009, the keynote address was titled, "Improving Achievement and Closing Gaps Between Groups: Lessons From Schools and Districts on the Performance Frontier." A decade later at NCSM's Annual Meeting in 2019, NCSM President Mona Toncheff observed that while we have made progress, we are asking the same questions alongside a host of new, deeper ones. She challenged the members of NCSM to take action now to ensure a changed discourse at the 2029 convening of NCSM.

Today, mathematics education leaders are flush with tools and strategies to equip our teachers to create meaningful mathematics learning experiences for their students. We have decades of research and better practices at our disposal with technology that makes those better routines of practice more accessible than ever before. In the face of an ever-changing national and global landscape, it has rarely been more necessary to employ *bold leadership* in order to ensure our students have equitable access to deep and meaningful mathematics. What does it take to be a bold mathematics leader? And what are the essential actions K–12 mathematics education leaders must employ each and every year to come closer to this vision of equity for learning?

Bold Mathematics Leaders Defined

You begin by unwrapping the term *bold mathematics leadership.* The word *bold* may be defined as "showing an ability to take risks; confident and courageous" (Google, n.d.) or as "involving a risk" (Macmillan, n.d.). When you lead in a bold way, you will need to take risks. When you have student groups that are underrepresented in fourth year mathematics courses or are placed in remedial mathematics courses with no plans for granting access to meaningful and relevant content, you need to be bold. When you walk into four classrooms at the same school and the same grade, and the mathematics lessons and expectations for student learning demonstrate an inconsistent rigor, there are wide discrepancies with teacher practice, or perhaps wrong routines for increasing student success, you need to be bold. The bold leadership actions described in the *Framework for Leadership in Mathematics Education* provide the focus for the knowledge and skills you must pursue in order to be bold.

Bold mathematics leaders **advocate** for and expect high-quality, equitable mathematics teaching and learning for each and every student. They **design** and implement structures that support the requisite equitable and high-quality mathematics teaching and learning by every teacher and team. Bold leaders **empower** and nurture a culture of productive professionalism and collective teacher efficacy (Bandura, 1996) in which all stakeholders embrace ways to continually improve their work. Bold leaders **monitor** instructional programs, collect data, provide feedback, and expect teachers to act on evidence (or lack thereof) of student learning. Bold leaders hold high standards for both themselves and their staff and expect their staff and faculty to hold students to those same high standards.

Ultimately, your success as a mathematics education leader resides in the impact you leave on the next generation of mathematics education teachers, leaders, and programs. When you decide to make a difference and make an impact, you choose to take a stand and commit others to a complex, yet crystal-clear set of leadership actions.

Bold leaders consider their guiding principles and take a stand in ways that support enacting those principles. The NCSM *Framework for Leadership in Mathematics Education* identifies 12 key essential actions. These actions are referenced as imperatives because they are crucial and of vital importance, and they help you to identify ways in which you can take a stand in support of teachers and students in your school organization. Every leader is capable of making commitments for things that matter and providing meaning to values that will significantly impact student learning.

This is what it means to live a courageous and ***bold*** leadership life.

Who Are Bold Mathematics Leaders?

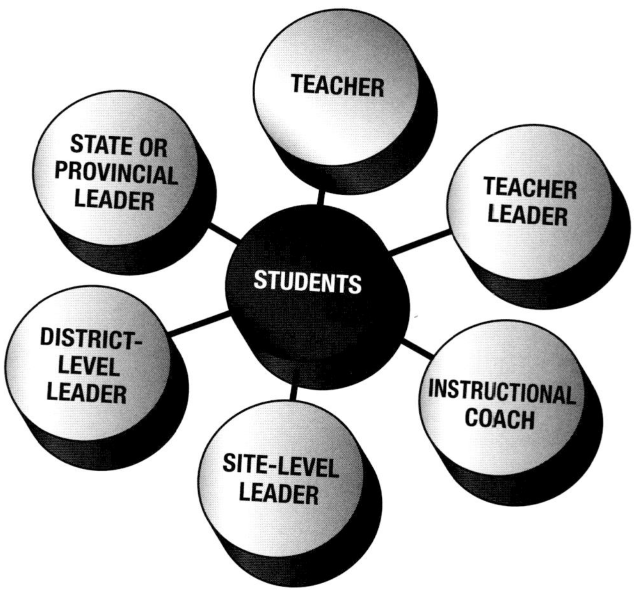

Figure I.1. Mathematics Education Leadership Roles

There are many different leadership roles in which you can serve in the mathematics education community, and each of these roles is important for students and teachers. For the purpose of the Essential Actions series, we have defined six mathematics education leadership roles (Figure I.1). Of these six, the role of teacher is unique as the educator who leads in her or his classroom and primarily interacts directly with students. We recognize the power of teachers as leaders and focus on five roles of mathematics education leadership: teacher leaders, instructional coaches, site-level leaders, district-level leaders, and state/provincial-level leaders. Each of these role's actions is aligned to the framework and ultimately will support improving student achievement. Some of these roles are traditionally recognized, such as provincial/state-level leaders, district-level leaders, or site-level leaders. Teacher leaders and coaches make important contributions to leadership (Reeves, 2009). The work of teacher leaders falls into four categories: collaborating, advocating, modeling, and providing resources (Cheung, Reinhardt, Stone, & Little, 2018). Figure I.2 describes some of the actions that people serving in these roles may perform as they lead mathematics education programs. In this book, we think of mathematics programs not as a set of curriculum materials one might purchase from a publisher; rather, they are the set of curriculum, instruction, and assessments necessary for the teaching and learning of mathematics.

Mathematics Leadership Role	Possible Actions for the Specific Role
Teacher Leader	• Orchestrate and facilitate the work of a grade-level or content-level mathematics team. • Assist the grade-level or course-based math team with making decisions and reaching agreements. • Serve as a main point person to request support from a coach or site leaders when challenges or questions arise with the mathematics instructional program. • Engage the community in learning and understanding the way mathematics is taught today.
Coach	• Collaborate with site leaders and/or district leaders to analyze current systemic structures and determine supports needed to improve a shared vision of mathematics teaching and learning. • Support the mathematics team leaders and members with the preconference, side-by-side coaching, and the post-conference to improve long-term teacher practice. • Provide job-embedded professional learning through workshops and engagement in intentional coaching cycles of planning, side-by-side co-teaching, and debriefing. • Collect and analyze data from multiple sources to inform and improve classroom practices. • Engage the community in learning and understanding the way mathematics is taught today.
Site Leader	• Provide feedback to teachers, teams, and leaders. • Grow in your own learning alongside the mathematics team leaders and coaches. • Provide non-evaluative support in a coaching manner to mathematics teachers and teams. • Facilitate the development of a campus-based vision for effective and equitable mathematics instruction. • Collect and analyze data from multiple sources to inform and improve campus programs. • Engage the community in learning and understanding the way mathematics is taught today.
District Leader	• Facilitate the development of a district vision for effective and equitable mathematics instruction. • Develop and promote district-wide goals. • Develop and implement district-level curriculum and instruction documents that directly address the needs of your population of students. • Ensure teachers are provided necessary resources to implement district-level curriculum (e.g., dual-language options, mathematics manipulatives). • Engage the community in learning and understanding the way mathematics is taught today. • Provide time for teachers, coaches, and site leaders to deepen their learning of district-level curriculum and best mathematics instructional practices. • Collect and analyze data from multiple sources to inform and improve district programs.
State/Provincial Leader	• Advocate for effective and equitable mathematics education policies. • Implement statewide or province-wide policies with an eye on equity. • Collect and analyze data from multiple sources to inform and improve province or state programs. • Engage the community in learning and understanding the way mathematics is taught today.

Figure I.2 The Five Roles of Mathematics Education Leaders

As you read Figure I.2, reflect on which role best describes the type of work you perform in your school or organization. Throughout the book, we will reference specific actions aligned to each role. The *Framework for Leadership in Mathematics Education* is designed to help you move forward in your journey to become a bold mathematics education leader.

On the Shoulders of Giants

In order to place the *Framework for Leadership in Mathematics Education* in a proper context, consider the foundational role played by two particular NCSM publications: *The PRIME Leadership Framework: Principles and Indicators for Mathematics Education Leaders* (NCSM, 2008) and *It's TIME: Themes and Imperatives for Mathematics Education* (NCSM, 2014b). These two landmark publications served the mathematics community well during the years of mathematics reform that brought the Common Core State Standards for Mathematics (CCSSO & NGA, 2010) and others states' college and career readiness-based standards into the curriculum of PreK–12 schools and classrooms. Each of these documents contains important theoretical underpinnings upon which the *Framework for Leadership in Mathematics Education* is constructed. They are, indeed, two giants upon whose shoulders this work stands.

Principles and Indicators for Mathematics Education (PRIME)

In 2008, NCSM published *The PRIME Leadership Framework: Principles and Indicators for Mathematics Education Leaders* (NCSM, 2008). The release of *PRIME* served as a bellwether event for mathematics leaders across North America. For the first time, leaders of Pre-K–12 mathematics were grounded in common language that described the vision of *what* leaders of effective Pre-K–12 mathematics programs should do.

Three assumptions created a lens through which the *PRIME* leadership framework viewed leadership responsibility.

1. Success for every student, teacher, and leader
2. Research-informed teacher actions
3. Teacher collaboration and professional learning

With these assumptions in mind, the *PRIME* leadership framework described four leadership principles and three indicators for each leadership principle, shown in Figure I.3. These indicators provided specific teacher actions for which leaders were responsible for cultivating.

Equity Leadership *Ensure high expectations and access to meaningful mathematics learning for every student.*	• Every teacher addresses gaps in mathematics achievement expectations for all student populations. • Every teacher provides each student access to relevant and meaningful mathematics experiences. • Every teacher works interdependently in a collaborative learning community to erase inequities in student learning.
Teaching and Learning Leadership *Ensure high expectations and access to meaningful mathematics instruction every day.*	• Every teacher pursues the successful learning of mathematics for every student. • Every teacher implements research-informed best practices and uses effective instructional planning and teaching strategies. • Every teacher participates in continuous and meaningful mathematics professional development and learning in order to improve his or her practice.
Curriculum Leadership *Ensure relevant and meaningful mathematics in every lesson.*	• Every teacher implements the local curriculum and uses instructional resources that are coherent and reflect state standards and national curriculum recommendations. • Every teacher implements a curriculum that is focused on relevant and meaningful mathematics. • Every teacher implements the intended curriculum with needed intervention and makes certain it is attained by every student.
Assessment Leadership *Ensure timely, accurate monitoring of student learning and adjustment of teacher instruction for improved student learning.*	• Every teacher uses student assessments that are congruent and aligned by grade level or course content. • Every teacher uses formative assessment processes to inform teacher practice and student learning. • Every teacher uses summative assessment data to evaluate mathematics grade-level, course, and program effectiveness.

Figure I.3. The *PRIME* Leadership Framework
Source: NCSM (2008)

Recognizing that leadership, similar to teaching and learning, is a journey, the *PRIME* leadership framework introduced the notion of stages for leadership growth within the mathematics community, shown in Figure I.4.

The three stages of leadership conceptualize how you grow and evolve over time with respect to each stage. In Stage 1 leadership, a leader builds self-awareness of emerging leadership and is recognized for her or his own teaching abilities. In Stage 2 leadership, a leader builds a collaborative network of colleagues and is seen by those colleagues as possessing the interpersonal skills and commitment for leading change within that network. In Stage 3 leadership, a leader engages in sustained, systemic implementation of each of the leadership indicators, earning respect for his or her influence in the larger education community outside of her or his immediate context.

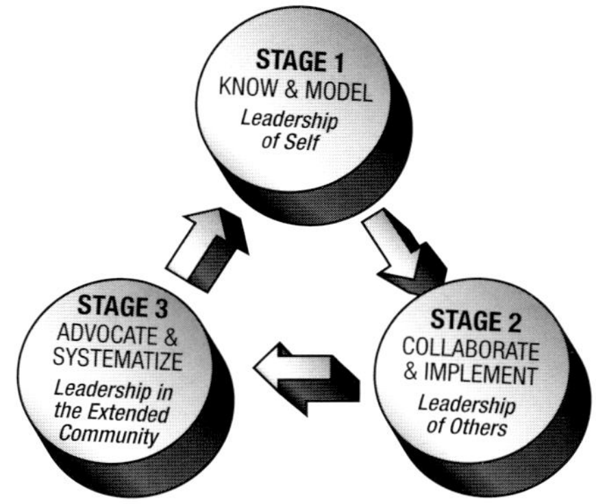

Figure I.4. Three Stages of Mathematics Leadership
Source: Adapted NCSM (2008).

PRIME challenged mathematics education leaders to employ what we now consider to be *bold leadership:* "Mathematics programs will only get better when leaders open themselves and other teachers to new ideas, risk imaginatively, and enthusiastically inspire those they lead with a desire to learn and grow together" (NCSM, 2008, p. 56). By employing the *PRIME* leadership framework, mathematics education leaders would be empowered to close the "knowing-doing" gap between what we know about enhancing student achievement and taking action resulting from that knowledge. The *PRIME* framework set the stage for leadership development by providing the "what" we need to do in order to work toward achieving our visions for equitable, high-quality mathematics education. We were invited to explore *how* to achieve those goals when *It's TIME* was released.

It's TIME: A Response to College and Career Readiness-Based Curriculum Standards

In 2014, NCSM released the companion document, *It's TIME: Themes and Imperatives for Mathematics Education.* The second publication from NCSM provided leaders with "how" they might go about accomplishing the lofty expectations set forth in *PRIME* (NCSM, 2014b). The overarching themes of *social justice, systemic thoughts and actions,* and *leadership* were combined with the imperatives for *knowledge, instruction and assessment,* and *systemic change. It's TIME* moved leaders toward the effective, consistent, and impactful implementation of the spirit of common mathematics curriculum standards connected to significantly higher levels of student achievement for those common Pre-K–12 mathematics standards.

It's TIME was developed to assist with the implementation of the Common Core State Standards for Mathematics (CCSSM); however, its lessons can be applied to any state standards that call for rigorous and relevant curriculum and instruction.

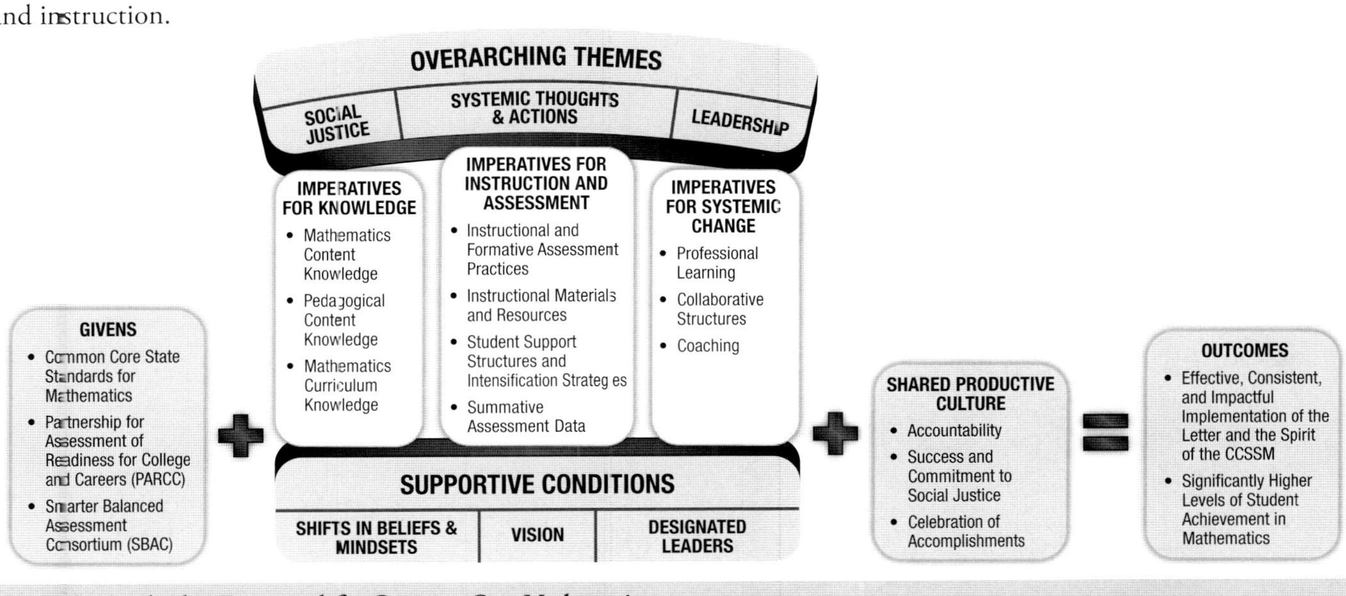

Figure I.5 Leadership Framework for Common Core Mathematics
Source: NCSM (2014b).

What made *It's TIME* particularly powerful was its specificity for *how* leaders could obtain the what outlined in the *PRIME* leadership framework.

For instance, one of the *PRIME* indicators (NCSM, 2008) in the Teaching and Learning principle states, "Every teacher participates in continuous and meaningful mathematics professional development and learning in order to improve his or her practice." *It's TIME* clarified the Teaching and Learning *PRIME* indicator by addressing specific types of teacher knowledge: mathematics content knowledge, pedagogical content knowledge, and mathematics curriculum knowledge. For teachers to effectively implement rigorous standards that lead students to appropriate levels of college and career readiness, such as CCSSM, leaders were expected to cultivate all three types of knowledge. This cultivation became *how* teachers engaged in continuous and meaningful learning.

It's TIME provided mathematics education leaders with specific steps that were necessary to implement curriculum standards that emphasized college and career readiness, specifically, the Common Core State Standards for Mathematics. Curriculum standards, however, go through periodic replacement and revision cycles.

Periodically, mathematics educators and leaders find themselves amid one of those cycles. The powerful ideas from *PRIME* and *It's TIME* remain as relevant today as when they were first presented. However, mathematics leaders today need more. Even with revised standards, increased accountability, and more targeted support for diverse learners, we, as a mathematics community, are not there yet. The *Framework for Leadership in Mathematics Education,* supported by current research and equitable instructional practices, describes how to be a bold mathematics leader to lead effective, equitable mathematics education programs.

Essential Actions: A Framework for Leadership in Mathematics Education

A guiding principle is a belief or ideal that defines the work of an organization and transcends changes in circumstance, leadership, or any of the other working parts of that organization. Through that lens, there are four guiding principles identified in the *Framework for Leadership in Mathematics Education.* The guiding principles include four key elements of any mathematics education leader's work and provide clarity and focus for how to lead Pre-K–12 mathematics. Mathematics education leaders will:

- **Advocate** and expect high-quality, equitable mathematics teaching and learning for every student
- **Design** and implement structures that support high-quality mathematics teaching and learning by every teacher
- **Empower** and nurture a culture of productive professionalism
- **Monitor** and act on evidence of student learning

As well, the guiding principles provide structure for defining the role of any individuals who find themselves leading mathematics instructional programs in a school, district, or institute of higher education.

We define an *imperative* as *"an essential or urgent action or idea that must be considered in effective mathematics education leadership."* Three specific imperatives, based on three stages of leadership that were illuminated in the original *PRIME* leadership framework (NCSM, 2008), frame each of the four guiding principles.

- Stage 1: Leadership of self
- Stage 2: Leadership of others
- Stage 3: Leadership beyond the local workplace

PRIME situated three stages of leadership in a trajectory so that leadership began with self, expanded to others, and then grew beyond the leader's immediate sphere of influence. Viewing leadership growth in this way allows a mathematics education leader to identify where she or he is along a leadership pathway and then chart a course for ongoing learning and leadership development in order to grow as a leader.

For the *Framework for Leadership in Mathematics Education,* we turn to the essential *actions* that mathematics education leaders, regardless of their role, may take in order to influence change. Through that lens, the focus shifts from *stages of leadership* to *commitment to action.* Thus, we reframe the *PRIME* imperatives of stages of mathematics education leadership to three imperatives of commitment to action (Figure I.6).

PRIME LEADERSHIP FRAMEWORK		ESSENTIAL ACTIONS FRAMEWORK FOR LEADERSHIP IN MATHEMATICS EDUCATION
STAGE 1: LEADERSHIP OF SELF Leadership of self-knowledge, awareness, development, and modeling of the 12 leadership indicators; the leader is respected for his or her own teaching and learning skills. This is the "know and model" stage of leadership growth and development.	→	**IMPERATIVE 1: COMMITMENT TO SELF** Commitments to action that an individual can take to lead those in the leader's immediate sphere of influence.
STAGE 2: LEADERSHIP OF OTHERS Leadership of all students and teachers within the mathematics program; leadership and development of other teachers, teams of teachers, and administrators toward full knowledge and development of each of the 12 leadership indicators; the leader is respected for his or her interpersonal skills and commitment for leading change among teams of teachers and colleagues. This is the "collaborate and implement" stage of leadership development.	→	**IMPERATIVE 2: COMMITMENT TO COLLEAGUES** Commitments to action that advance beyond the leader's immediate sphere of influence to build systemic structures and relationships.
STAGE 3: LEADERSHIP IN THE EXTENDED COMMUNITY Leadership of district, state, province, or beyond reform efforts through sustained deepened systemic implementation of each of the 12 leadership indicators. The leader is respected for his or her influence and engagement with an expanded community of educational stakeholders. This is the "advocate and systematize" stage of leadership development.	→	**IMPERATIVE 3: COMMITMENT TO OTHERS** Commitments to action that extend beyond the school into the community in which the school is situated.

Figure I.6. Shifts From Stages of Leadership to Commitments to Action

Re-framing the leadership imperatives to commitments to action allows us to build on what we know about effective leadership and translate it into actions that leaders may take to work toward the shared vision of high-quality instructional programs.

The combination of four guiding principles and three imperatives creates 12 essential actions that complete the framework (Figure I.7). These essential actions emerge from a careful study of research and better practices that support improved student learning and increased student achievement. When done looking through the lens of equity, mathematics leaders pursuing these actions can ensure each and every learner experiences meaningful and relevant mathematics.

Guiding Principle	Imperative 1 *Commitment to Self*	Imperative 2 *Commitment to Colleagues*	Imperative 3 *Commitment to Others*
ADVOCATE and expect high-quality, equitable mathematics teaching and learning for every student.	*Ensure that every teacher possesses a shared understanding and vision of high-quality mathematics instruction and the actions required to meet the vision.*	*Ensure that every teacher possesses the skills and knowledge necessary to design and implement meaningful learning experiences that lead to student understanding of mathematics.*	*Ensure that all stakeholders have a clear understanding of high-quality mathematics instruction and how to support it.*
DESIGN and implement structures that support high-quality mathematics teaching and learning for every teacher.	*Ensure mathematics learning for all students through organizational structures, time and resource allocation, and systemic supports that are aligned, intentional, and equitable.*	*Ensure systems of continual collaborative, job-embedded professional learning to build teacher and leader capacity and increase efficacy.*	*Ensure sustainability through engaging all stakeholders in systemic, long-range strategic planning for all teaching and learning improvement initiatives.*
EMPOWER and nurture a culture of productive professionalism.	*Ensure assumptions, beliefs, expectations, and habits are examined in order to shape the school or department culture around teaching and learning of mathematics.*	*Ensure a culture of reflection, refinement, and action focused on continuous improvement in mathematical learning.*	*Ensure students, teachers, families, and community partnerships are built upon meaningful engagement.*
MONITOR and act on evidence of student learning.	*Ensure the design and use of high-quality, aligned assessments and equitable assessment processes that guide meaningful reflection and action.*	*Ensure that the evidence of learning collected from every assessment is used to inform the design of curriculum, instruction, and the assessments themselves.*	*Ensure every student is provided access to grade-level content and intensification based on evidence of student learning.*

Figure I.7. Framework for Leadership in Mathematics Education

The framework's four guiding principles and 12 essential actions, which will be unwrapped throughout the rest of the book, are grounded in four foundational assumptions of our work life and are aligned to the mission and vision of NCSM. These foundational elements are:

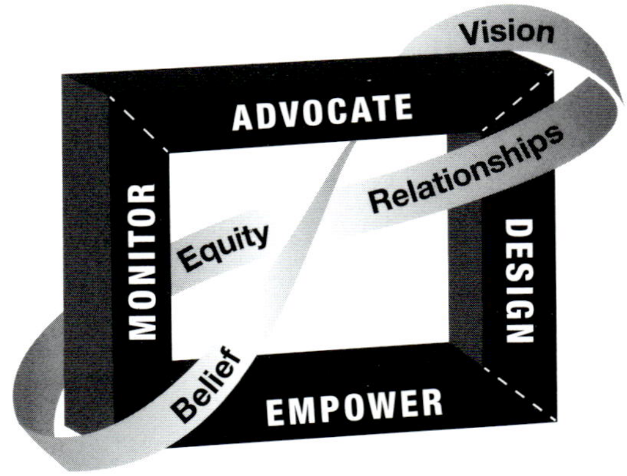

- **Beliefs:** The core belief of NCSM, and underlying assumption of its work, is to ensure all students experience and learn meaningful and relevant mathematics. In order to meet the needs of all learners, it is a mathematics education leader's responsibility to examine and facilitate alignment of all stakeholders' beliefs regarding mathematics education.
- **Vision:** Change is essential to improving students' mathematical learning, and this change cannot occur unless all stakeholders have a shared vision of high-quality, research-affirmed teaching that leads to all students learning meaningful and relevant mathematics.
- **Equity:** All established systems and structures ensure that all students have access to meaningful and relevant mathematics teaching and learning.
- **Relationships:** To implement and achieve a vision for mathematics teaching and learning for all, we work collectively and build working relationships among all stakeholders.

Your "bold leadership" requires a risk-taking, courageous, and confident mindset. The essential actions described in this framework support your bold mathematics leadership by outlining specific actions you can take to create and sustain equitable mathematics educational programs. Since these essential actions are grounded in research and equitable instructional practices, they provide you the safety to take the necessary risks in moving your teachers and colleagues, as well as their programs, forward for the specific purpose of improved student learning in mathematics.

How Do I Use the Framework?

Thomas Sergiovanni (1992) describes three aspects of leadership: the heart, head, and hand. The *heart* of leadership connects to what the leader values and believes: the leader's personal vision. The *head* of leadership connects to the leader's mindscape, or theory of practice, of how the world works, including reflection on situations that leaders may face. The *hand* of leadership connects to the leader's decisions, actions, and behaviors. The interactions among these three aspects of leadership create a structure through which we can think about how what we know about leadership and learning (the head of leadership) influences what we believe about leadership and learning (the heart of leadership) and drives our actions (the hand of leadership). The *Framework for Leadership in Mathematics Education* presents you with an opportunity to examine all three of these aspects as they relate to your work and your particular leadership role.

In creating the *PRIME Leadership Framework,* NCSM (2008) acknowledged that "a single mathematics education leader can have an incredible impact on the development and effectiveness of others" (p. 1). Reeves (2004), in his extensive work with education leadership, observed that "in the vast majority of leadership evaluation documents I reviewed, one of two problems prevailed. Either the standards themselves were ambiguous, or performance expectations were unclear" (p. 5). We also know that "clear is kind; unclear is unkind" (Brown, 2018, p. 48). Hence, the *Framework for Leadership in Mathematics Education* presents four guiding principles tied to 12 specific essential actions that you can take in your leadership role to improve how students in your school, district, or organization learn mathematics.

We have organized this framework around the four guiding principles. Each chapter takes you through several big ideas related to that guiding principle and digs deeper into how those ideas manifest within each imperative, as shown in Figure I.8.

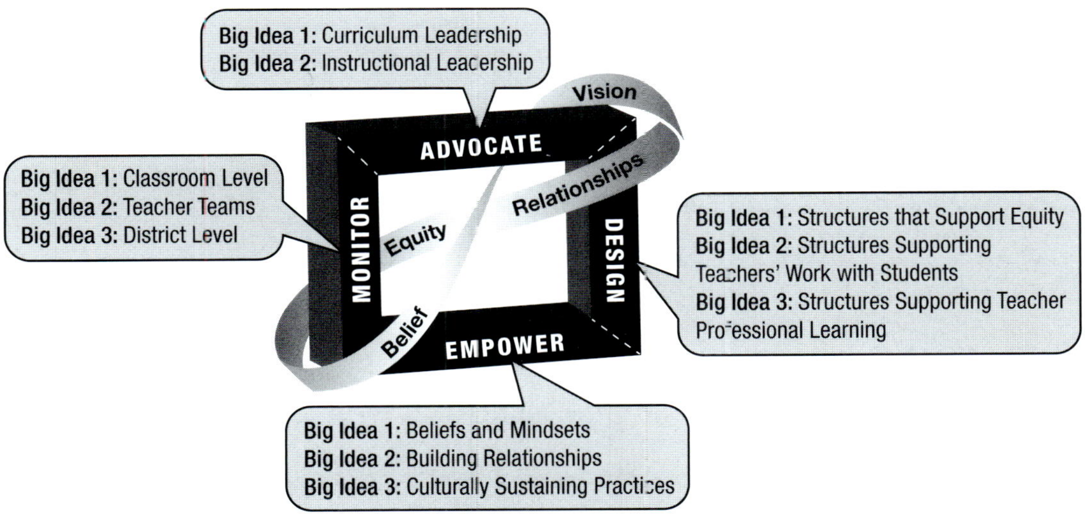

Figure I.8. Big Ideas for Each Guiding Principle

Each chapter examines the big ideas listed in Figure I.8 that effective mathematics education leaders take with respect to a particular guiding principle. You will see how each big idea evolves through the three imperatives: *commitment to self, commitment to colleagues,* and *commitment to others.*

In Chapter 1, you will explore the first guiding principle: *ADVOCATE and expect high-quality, equitable mathematics teaching and learning for every student.* This chapter will provide you with insights on how you, as a mathematics education leader, might leverage your leadership to advocate for instructional programs that support student success. You will learn more about three essential actions relating to the big ideas of curriculum leadership and instructional leadership.

- *Ensure that every teacher possesses a shared understanding and vision of high-quality mathematics instruction and the actions required to meet the vision.*
- *Ensure that every teacher possesses the skills and knowledge necessary to design and implement meaningful learning experiences that lead to student understanding of mathematics.*
- *Ensure that all stakeholders have a clear understanding of high-quality mathematics instruction and how to support it*

In Chapter 2, you will investigate the second guiding principle: *DESIGN and implement structures that support high-quality mathematics teaching and learning for every teacher.* This chapter will guide you through processes that you can use as you design instructional programs and implement specific structures that support teachers. You will learn more about three essential actions relating to the big ideas of designing structures that support teaching and learning, structures supporting professional learning, and structures supporting evaluation.

- *Ensure mathematics learning for all students through organizational structures, time and resource allocation, and systemic supports that are aligned, intentional, and equitable.*
- *Ensure systems of continual collaborative, job-embedded professional learning to build teacher and leader capacity and increase efficacy.*
- *Ensure sustainability through engaging all stakeholders in systemic, long-range strategic planning for all teaching and learning improvement initiatives.*

In Chapter 3, you will uncover the third guiding principle: *EMPOWER and nurture a culture of productive professionalism.* This chapter will discuss ways that you can empower your colleagues to grow as educators and foster a culture of productive professionalism. You will learn more about three essential actions relating to the big ideas of beliefs and mindsets, building relationships, and culturally sustaining practices.

- *Ensure assumptions, beliefs, expectations, and habits are examined in order to shape the school or department culture around teaching and learning of mathematics.*
- *Ensure a culture of reflection, refinement, and action focused on continuous improvement in mathematical learning.*
- *Ensure students, teachers, families, and community partnerships are built upon meaningful engagement.*

In Chapter 4, you will explore the fourth and final guiding principle: *MONITOR and act on evidence of student learning.* This chapter will provide an overview on processes that you can use to collect data and use that data to inform your work as a leader. You will learn more about three essential actions relating to the big ideas of monitoring student learning at the classroom level, monitoring student learning at the site level, and monitoring student learning at the district level.

- *Ensure the design and use of high-quality, aligned assessments and equitable assessment processes that guide meaningful reflection and action.*
- *Ensure that the evidence of learning collected from every assessment is used to inform the design of curriculum, instruction, and the assessments themselves.*
- *Ensure every student is provided access to grade-level content and intensification based on evidence of student learning.*

As you read through the chapters, you will have opportunities to reflect on how the ideas presented relate to your role as a mathematics education leader. Additionally, we provide tools you can use with your colleagues as you lead them through the work of the essential actions. We encourage you to use these tools and self-reflections as you consider how to implement the framework into your current leadership work. When you **advocate** for high-quality and equitable teaching and learning, **design** and implement structures that support high-quality teaching and learning, **empower** and nurture a culture of productive professionalism, and **monitor** and use evidence of learning to inform instructional and curriculum decisions, with a laser-sharp focus on equity for all students, you move closer to achieving a vision of mathematics teaching and learning for all students.

Additionally, we provide some resources online that support your work as you implement ideas from this book. Those resources are available online and we use the mouse icon to indicate them.

CHAPTER 1

Guiding Principle 1: Advocate

Big Idea 1: Curriculum Leadership for High-Quality, Equitable Mathematics Teaching and Learning

Big Idea 2: Instructional Leadership for High-Quality, Equitable Mathematics Teaching and Learning

Summary of Essential Actions

Taking Action

"Advocate for each and every student to have access to rigorous mathematics that develops their understanding, skills, and knowledge, along with the confidence to leverage their learning, in order to improve their world."

—NCSM Vision Statement

GUIDING PRINCIPLES

ADVOCATE and expect high-quality, equitable mathematics teaching and learning for every student.

DESIGN and implement structures that support high-quality mathematics teaching and learning for every teacher.

EMPOWER and nurture a culture of productive professionalism.

MONITOR and act on evidence of student learning.

Guiding Principle 1: ADVOCATE and expect high-quality, equitable mathematics teaching and learning for every student.

The first guiding principle provides clarity for the work of mathematics education leaders, requiring each type of leader to **advocate** for and expect high-quality, equitable mathematics teaching and learning for every student. The operative word in this principle is *every* student, including the students who need additional support to access high-quality, equitable mathematics teaching. Consider the scenarios in Figure 1.1.

Scenario 1: A team of sixth-grade teachers with representatives from each campus in a school district meets during the summer for 1 week to create curriculum documents for their district to use. These documents will frame the scope and sequence of instruction and provide teachers with instructional resources from which they can select to use with their students.
Scenario 2: A third-grade teacher team meets regularly in a professional learning community setting. As they prepare for an upcoming unit, they will investigate which instructional strategy best matches the learning goal for a particular topic in that unit.

Figure 1.1. Advocate and Expect High-Quality, Equitable Mathematics Teaching and Learning Scenarios

The first scenario portrays teachers working collaboratively in pursuit of a shared goal: creating curriculum documents they and their colleagues can use to guide their work throughout the academic year. Collaboration is essential to a productive professional culture (see Chapter 3) and is critical to fostering a teaching culture that espouses continuing learning (DuFour & Eaker, 1998). However, how does the leader know that the teacher team is using high-quality resources to build the curriculum documents? How do the curriculum documents help teachers move student learning toward the district's or site's espoused vision?

The second scenario shows a teacher team working collaboratively to align instruction with stated learning goals. Collaborative structures such as a grade-level teacher team are important elements of a school culture. However, what criteria are teachers using to evaluate instructional strategies? How do they know the strategies will reach all students? How do they know that the instructional strategy, instructional task, and learning goal all align and feature culturally relevant teaching?

Both of these scenarios showcase intentional structures that support teacher collaboration in pursuit of improved teaching and learning. Both of these scenarios indicate situations that are ripe for advocating for high-quality, equitable teaching, and learning. Effective leaders layer in additional supports to ensure that the collaboration yields results that align with the shared vision for the campus or district and are rooted in what research tells us about high-quality teaching and learning for all students.

To advocate for and expect high-quality, equitable mathematics teaching and learning requires mathematics leaders to have a deep understanding of the integrated nature of the elements that drive the mathematics teaching and learning process: curriculum, instruction, and assessment. During this chapter, we look at the role that curriculum and instruction plays in defining the opportunities and experiences for students to access high-quality, rich, and relevant mathematics with a focus on the actions leaders can take to ensure that the Imperatives, shown in Figure 1.2, come to fruition. We will address the third element, assessment, in Chapter 4 within the fourth guiding principle: *MONITOR and act on evidence of student learning.*

	Imperative 1 *Commitment to Self*	**Imperative 2** *Commitment to Colleagues*	**Imperative 3** *Commitment to Others*
ADVOCATE and expect high-quality, equitable mathematics teaching and learning for every student.	**Essential Action** *Ensure that every teacher possesses a shared understanding and vision of high-quality mathematics instruction and the actions required to meet the vision.*	**Essential Action** *Ensure that every teacher possesses the skills and knowledge necessary to design and implement meaningful learning experiences that lead to student understanding of mathematics.*	**Essential Action** *Ensure that all stakeholders have a clear understanding of high-quality mathematics instruction and how to support it.*

Figure 1.2. Advocate Guiding Principle From the *Framework for Leadership in Mathematics Education*

Within the advocate guiding principle, the commitments to action among the different imperatives emerge as follows:

- Actions related to the first imperative, *commitment to self,* focus on how a leader works with educators in her or his sphere of influence to create a shared vision of high-quality teaching and learning.
- Actions related to the second imperative, *commitment to colleagues,* focus on how a leader works with educators in adjacent spheres of influence, such as with other teacher teams, with other departments, on a campus, or within the district, to ensure that all teachers possess the knowledge and skills necessary to implement meaningful learning experiences for students.
- Actions related to the third imperative, *commitment to others,* focus on how a leader works outside of his or her immediate learning community in order to build a common understanding among all stakeholders of what high-quality mathematics instruction looks like.

The **advocate** imperatives demonstrate the connections among every teacher having a shared vision of what high-quality, equitable mathematics instruction is, every teacher possessing the skills and knowledge necessary to teach toward that vision, and all stakeholders having a clear understanding of that shared vision and how it benefits all members of the community. This chapter is designed to help bring these imperatives to action by focusing on two big ideas:

- Big Idea 1: Curriculum leadership for high-quality, equitable mathematics teaching and learning
- Big Idea 2: Instructional leadership for high-quality, equitable mathematics teaching and learning

As a mathematics leader, what is your role in advocating and expecting high-quality, equitable mathematics teaching and learning for every student? Use the guiding questions in Figure 1.3 to reflect, based on your leadership role, on how you as a bold leader can support the three imperatives within the **Advocate** guiding principle.

Guiding Principle 1: ADVOCATE and expect high-quality, equitable mathematics teaching and learning for every student.

Essential Action	Guiding Questions
Ensure that every teacher possesses a shared understanding and vision of high-quality mathematics instruction and the actions required to meet the vision.	In my leadership role: • Do we have a shared vision for mathematics teaching and learning? • How well do my colleagues and I understand and implement the effective mathematics teaching practices to meet the vision?
Ensure that every teacher possesses the skills and knowledge necessary to design and implement meaningful learning experiences that lead to student understanding of mathematics.	In my leadership role: • How do we discuss meaningful and relevant mathematics? • How do we create learning experiences that lead to increased student learning? • How do we include culturally relevant tasks in our curriculum?
Ensure that all stakeholders have a clear understanding of high-quality mathematics instruction and how to support it.	In my leadership role: • How do we build collective capacity with high-quality mathematics instruction? • How do we communicate our vision of mathematics teaching and learning to families and community organizations? • How do we promote culturally relevant learning experiences?

Figure 1.3. Leadership Roles With the Advocate Guiding Principle

Visit http://mathedleadership.org/EAresources for a role-specific reflection tool with questions for each mathematics leadership role (teacher leader, coach, site-based leader, district leader, and state/provincial leader).

As we explore the **Advocate** guiding principle, consider your reflection in Figure 1.3 to plan next steps.

Big Idea 1: Curriculum Leadership for High-Quality, Equitable Mathematics Teaching and Learning

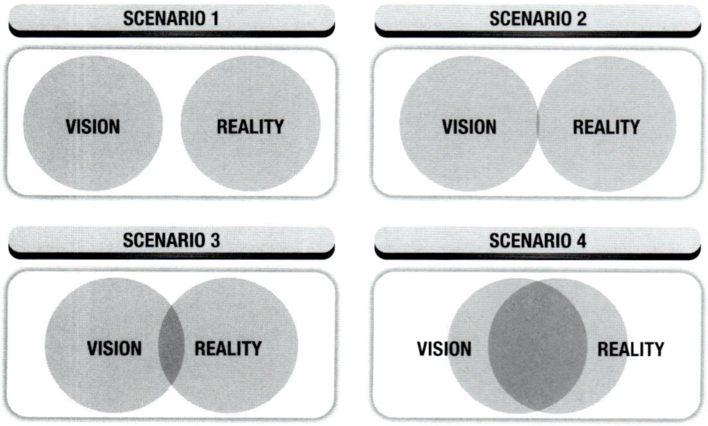

Figure 1.4. Relationships Between Vision (Intended Curriculum) and Reality (Implemented Curriculum)

Consider the four scenarios shown in Figure 1.4. Which one best represents the alignment between the vision for the intended curriculum that is designed and planned by teacher teams at the campus or district level and what is actually observed in each classroom at your school or across your district, state, or province?

Acknowledging the relationship between the intended curriculum and the implemented curriculum, or the reality that is lived out in mathematics classrooms, helps leaders determine a particular course of action to improve that alignment.

Clear Shared Vision of Curriculum

Essential Action

Ensure that every teacher possesses a shared understanding and vision of high-quality mathematics instruction and the actions required to meet the vision.

The imperative of *commitment to self* includes taking actions directly related to supporting teachers and educators in your care or in your immediate sphere of influence. Regarding curriculum leadership, the initial responsibility is to work collaboratively to create a shared vision of high-quality curriculum.

Schools typically display "*mission* statements that describe their purpose and vision statements that describe what they hope to become" (NCSM, 2019b, p. 25). A strong vision for mathematics teaching and learning provides guidance for teachers, leaders, and other stakeholders to work together to accomplish a common goal. Having a shared vision is important because all stakeholders then understand *why* their actions are important toward accomplishing the common goal. Creating a strong shared vision "requires collaboration and a common understanding of the actions that will support meeting the vision" (NCSM, 2019b, p. 25). Collaboratively developing a shared vision for high-quality mathematics teaching and learning is important because we know that "people engaged in collaborative team learning are able to learn from one another, thus creating momentum to fuel continued improvement" (DuFour & Eaker, 1998, p. 27). Effective mathematics education leaders work to build collaboratively such a shared vision.

Use a protocol such as the Vision Casting Protocol (Figure 1.5) to guide discussions among your colleagues and create a shared vision of high-quality curriculum.

Ask your team the following: *What does an effective mathematics classroom and mathematics lesson look like and sound like to meet the needs of each and every learner?* Or, ask, *What does equitable instruction look like and sound like?*

Give each person a stack of sticky notes and have each person individually brainstorm the essential factors of their vision for instruction, per sticky note, for a specified amount of time. You might consider using two different-colored sticky notes with teams: One color represents the part of the vision that they feel they have truly worked toward as a team, and the other color represents a component of the vision that they feel that they need to still work toward.

Once the brainstorming is complete, the team sorts the sticky notes into the three to five big ideas of their vision and creates a vision statement on mathematics instruction that they all agree upon and will commit to working toward. For the teams that use two different colors, when the sticky notes are sorted they create a visual representation of their vision and actions that have been addressed as a team and the actions that the team still needs to address. The teams can create or revise current action steps to meet their vision. You can follow this same protocol for creating vision statements with assessment, grading and homework, and intervention.

Figure 1.5. Vision Casting Protocol
Source: Adapted from NCSM (2019b, p. 33).

Visit www.mathedleadership.org/EAresources to download a free reproducible version of this figure.

Once the vision is complete, leaders needs to work on creating curriculum that will support meeting the intended vision.

The Essential Role of Curriculum Leadership

Curriculum leadership begins with a full understanding of the mathematics program, particularly how it fits into the state/provincial and local expectations for developing students who are prepared and ready to pursue postsecondary education and training, start a career, and contribute to their community. In this book, we think of a *mathematics program* as the set of curriculum, instruction, and assessment elements that come together to create students' learning experience. States and provinces have each adopted a set of curriculum standards, which are a set of "learning goals for what students should know and be able to do at each grade level" or course (CCSSO & NGA, 2010). These provincial and state curriculum standards are then articulated into a district or school curriculum.

The importance of creating a shared vision of the intended mathematics curriculum increases with the adoption of more rigorous mathematics standards. Mathematics leaders must be intentional in defining, designing, and developing the intended curriculum, developed or adopted by and with input from all stakeholders so that it becomes the taught and assessed curriculum and thus the guaranteed and viable curriculum that each and every student learns. A *guaranteed and viable curriculum* is one that gives students access to the same essential learning regardless of who teaches the class, and it allows adequate time to teach the content (Barnes & Toncheff, 2016; Marzano, 2003).

As a mathematics leader, one must advocate that the curriculum supports equitable access to meaningful mathematics for every student. This involves a commitment to transparency, engaging all stakeholders in the process of designing and developing a curriculum with common learning goals for each and every grade level or course and common learning pathways—where tracking is nonexistent or limited—and common expectations and outcomes.

The Essential Components of the Mathematics Curriculum

Mathematics curriculum is a developmental listing of knowledge and skills for which students should demonstrate mathematical competence—in content standards, proficiencies, processes, and practices. At the core of the mathematics curriculum lies the set of mathematical standards, often established by an educational agency, that all students are expected to learn in your state, province, or local school system as they progress from grade to grade. Many educational agencies have established the mathematical competencies based on recommendations from several key documents.

- National Council of Teachers of Mathematics:
    - *Principles and Standards for School Mathematics* (2000)
    - *Curriculum Focal Points for Prekindergarten Through Grade 8 Mathematics: A Quest for Coherence* (2006)
    - *Principles to Action: Ensuring Mathematical Success for All* (2014)
    - *Catalyzing Change for High School Mathematics: Initiating Critical Conversations* (2018)
- National Research Council (NRC): *Adding It Up: Helping Children Learn Mathematics* (2001)
- Common Core State Standards for Mathematics (CCSSO & NGA, 2010)

The clear articulation of the mathematics curriculum is an artifact that encompasses the standards that are both vertically and horizontally connected. The artifacts that are in print or electronic format and accessible to all teachers for study and use are crucial to ensuring that the intended curriculum becomes the taught and assessed curriculum. The mathematics content expectations for each grade or course should be organized into feasible teaching guides that link mathematical competencies and instructional resources that will help teachers implement the content standards (NCTM, 2014b). The components shown in Figure 1.6 should be included in the mathematics curriculum.

Components of a Strong Mathematics Curriculum

- A rationale that relates the goals of mathematics to the school or local school system mission and vision
- A general description for the mathematical competencies—content standards, proficiencies, processes, and practices
- Specific measurable learning goals that are focused, coherent, aligned to standards, and organized to build across the Pre-K–12 mathematics sequence
- Alignment of measurable metrics, progress indicators, and performance measures for each grade level or course to the knowledge, skills, and competencies students need to meet the school/system goals and state or provincial standards
- A core set of instructional tasks and sample lessons designed for a culturally diverse student population—respect and value students' background, family, culture, linguistics, and lived experiences
- Research-informed and equity-based instructional strategies that support all learners

Figure 1.6. Components of a Strong Mathematics Curriculum
Source: Adapted from NCSM's *PRIME Leadership Framework* (2008).

Use the Curriculum Components Analysis Tool (Figure 1.7) to determine how well your curriculum aligns to the components of a strong mathematics curriculum.

Read each statement and rate level of agreement with each statement.

1	2	3	4	5
Strongly Disagree	Disagree	Neutral	Agree	Strongly Agree

Rating	Statement
	Our curriculum documents have a clear rationale that relates the goals of the mathematics to the school or local school system's mission and vision.
	Our curriculum documents contain a general description for the mathematical competencies—content standards, proficiencies, processes, and practices.
	Our curriculum documents contain specific measurable learning goals that are focused, coherent, aligned to standards, and organized to build across the PreK–12 mathematics sequence.
	Our curriculum documents show an alignment of measurable metrics, progress indicators, and performance measures for each grade level or course to the knowledge, skills, and competencies students need to meet the school/system goals and state or provincial standards.
	Our curriculum contains a core set of instructional tasks and sample lessons designed for a culturally diverse student population—respect and value students' background, family, culture, linguistics, and lived experiences.
	Our curriculum and its supporting documents support research-informed and equity-based instructional strategies that support all learners.

Figure 1.7. Curriculum Components Analysis Tool

Visit www.mathedleadership.org/EAresources to download a free reproducible version of this figure.

A vision of high-quality mathematics teaching and learning will require mathematics leaders to exercise curriculum leadership as they develop and advocate for a shared common understanding of the intended curriculum for all stakeholders. The curriculum is a core artifact that supports the teaching and learning process and represents the beliefs and expectations for high-quality teaching and learning to provide equitable access for all students to rich, meaningful learning experiences.

As a mathematics leader, what is your role in developing a shared understanding of the intended mathematics curriculum? Consider the questions in Figure 1.8 related to your role.

Mathematics Leadership Role	Possible Actions for the Specific Role
Teacher Leader	• How will I support my peers with developing a strong understanding of the grade-level/course mathematical competencies? • How will I ensure my peers' voices are included in choosing and/or designing?
Coach	• What opportunities do I need to facilitate to deepen teachers' mathematics curriculum knowledge? • How can I model the use of curriculum materials in the design and implementation of lessons?
Site Leader	• How will I provide feedback on the implementation of the curriculum? • What opportunities will help all stakeholders gain an understanding of the mathematics curriculum?
District Leader	• How will I ensure parents and guardians gain an understanding of the mathematics curriculum, so that they can support their learner(s) appropriately? • How will I provide teachers and administrators time to develop a collective understanding of the curriculum? • How will we gather evidence on the implementation of the intended curriculum, so that we can make adjustments?
State/Provincial Leader	• How do I ensure that all schools/school systems have access to a curriculum that is aligned to the state/provincial expectations?

Figure 1.8. Leadership Role-Based Curriculum Reflection Questions

Visit www.mathedleadership.org/EAresources to download a free reproducible version of this figure.

Monitoring Curriculum Programs

Effective leaders of mathematics model self-reflection and evaluation by creating transparent program improvement processes. We are charged with not only building a common vision for mathematics teaching and learning but also implementing and monitoring progress toward that vision. These necessary processes anchor curriculum leadership within NCSM's core belief that a shared **vision** is essential for enacting change that will improve students' mathematical learning. Mathematics program improvement is cyclical, ongoing, and adaptable to the ever-evolving needs of schools and the community.

Effective leaders of mathematics create collaborative systems for scrutinizing grade-level and course curriculum. NCSM considers *curriculum* to include clearly identified learning standards, assessments that are designed to measure student progress toward those standards, the research-affirmed instructional strategies that serve as the vehicle for student learning, and research-affirmed supports to students who require enrichment and/or intervention.

The mathematics curriculum evaluation process sets out to answer the questions below. When considering these questions, note that *mathematics program* refers to mathematics course progressions, programs of students, or other learning opportunities.

- How effective are planned courses or programs of study in supporting the achievement of desired student learning outcomes?
- Is the mathematics program well articulated, cohesive (vertically), and coherent?
- To what degree is the mathematics program aligned to intended student learning outcomes?
- Is there a tight alignment among the intended, taught, and assessed curriculum?

There are a number of curriculum evaluation methodologies available to mathematics leaders. We recommend that leaders work with a team of vested stakeholders to identify a methodology and criteria to guide the work. Organizations such as the National School Reform Faculty (NSRF; www.nsrfharmony.org) provide a plethora of protocols to guide collaborative work. The Alignment Protocol, based on the NSRF's Tuning Protocol, is shown in Figure 1.9 and is particularly useful for evaluating the alignment of instructional tasks to a subset of curriculum standards using student work as evidence of what the task asks students to demonstrate.

The Alignment Protocol

Setup: Create groups of up to 15. Preconference with the presenter so that artifacts are available and the presenter has developed a focusing question that guides the protocol. The presenter will need sufficient copies of artifacts or technology to display them. Participants will need notetaking materials.

Steps:

1. **Setup** (5 min). Review the purpose and steps of the protocol so participants know what to expect. If this is the first time a group is experiencing the protocol, talk about the importance of giving warm and cool feedback and why the presenter moves away from the group in Step 6.
2. **Present** (10 min). Distribute the artifacts (e.g., student work samples). The presenter shares the context around the work, such as the learning goal associated with the work, pertinent curriculum standards, the presenter's desired outcomes, or anything the presenter believes does not work well. End by stating the focusing question that specifically gives the group parameters for their feedback. It is helpful for this question to be posted and visible for the group in the subsequent steps.
3. **Clarify** (10 min). Participants ask clarifying questions of the presenter. Clarifying questions are those with brief, factual answers and are used to help participants better understand the artifacts, intended purpose of the work, and focusing question.
4. **Examine** (5–10 min). Participants review the work through the lens of the focusing question. Participants should take notes for *warm feedback,* where the work seems to be aligned with the stated goals, and *cool feedback,* which is designed to strengthen the work. This step is performed silently.
5. **Separate** (1 min). The presenter withdraws from the conversation at this point. It is helpful for the presenter to physically move away from the group, remaining close enough to hear and avoiding eye contact with the group. The group will next conduct a conversation about the work while the presenter silently takes notes.
6. **Warm and cool feedback** (15 min). The facilitator should remind the group that warm feedback may include comments about how the work is aligned with desired goals, and cool feedback offers ideas or suggestions for strengthening the work by addressing the focusing question. It is important that cool feedback is given as a way to answer the focusing question so that it is most helpful to the presenter. Participants will share feedback with each other as though having a conversation while the presenter is silent and takes notes.
   Begin with warm feedback. Then transition into cool feedback around the focusing question, couched in "I wonder..." statements. If the feedback begins to stray, the facilitator may need to remind participants of the posted presenter's focusing question in order to be the most helpful.
7. **Reflect** (5 min). Invite the presenter to return to the group. The presenter need not defend or address every point in the group's conversation. The presenter may reflect aloud to what she or he heard, particularly feedback that was insightful or helpful, while the participants are silent.
8. **Debrief** (5–10 min). It is important for the group to discuss how the process of the protocol supported the collaborative work. Begin by asking the presenter questions such as the following:
   - *How was the experience of presenting?*
   - *What did you think of turning your chair away and listening silently?*
   - *How was the outcome of this protocol different from other attempts to improve the work?*

Then open to debrief with the rest of the group.

- *How did the experience feel from your point of view?*
- *Did anything the presenter said surprise you?*
- *What have you learned that you may use in your work or when facing your own dilemmas?*

Figure 1.9. Alignment Protocol
Source: Adapted from the National School Reform Faculty (NSRF) Tuning Protocol.

Visit www.mathedleadership.org/EAresources to download a free reproducible version of this figure.

NCSM published a Curriculum Analysis tool that can also be used to support teachers and teams engaging in analysis of their instructional resources. Visit www.mathedleadership.org/EAresources to download the Materials Analysis Tool.

Now that you have reflected on your role in developing a shared understanding of the intended mathematics curriculum, it is time to turn your attention to the essential actions you must take to ensure that all teachers you serve understand the mathematics curriculum.

Develop Mathematics Curriculum Knowledge

Essential Action

Ensure that every teacher possesses the skills and knowledge necessary to design and implement meaningful learning experiences that lead to student understanding of mathematics.

As a part of your commitment to actions supporting your colleagues, it is important that you advocate for time and opportunities for teachers to develop their mathematics curriculum knowledge. This is true for teachers in your immediate sphere of influence as well as teachers in nearby spheres within your system or organization. To close the gap between the intended and implemented curriculum, we must advocate for time for teachers and leaders to make sense of the curriculum. The opportunities must allow teachers to investigate the curriculum at their grade level and across grade levels to fully understand the curriculum they expect students to learn.

It is important that mathematics leaders provide ongoing professional learning opportunities and resources to develop teachers' curriculum knowledge that builds knowledge of

- curriculum design and instructional resources,
- curriculum coherence,
- sound lesson design,
- content, and
- relevant meaningful mathematics.
  (Barnes & Toncheff, 2016; NCSM, 2008)

Ball and Cohen (1996) stress the importance of *educative curriculum,* suggesting that curriculum resources could improve instruction through supporting implementation as an opportunity for teacher professional learning. Furthermore, Lynch, Hill, Gonzalez, and Pollard (2019) found that STEM programs saw stronger outcomes when they helped teachers learn to use curriculum materials; focused on improving teachers' content knowledge, pedagogical content knowledge, and/or understanding of how students learn; incorporated summer workshops; and included teacher meetings to troubleshoot and discuss classroom implementation.

Providing grade-level or course teams regular time to plan with the curriculum in a professional learning setting allows teachers to gain an understanding of how and why focus, depth, and coherence make a mathematics curriculum effective. Collaborative planning also provides teachers and teacher teams the opportunity to identify and act on inconsistencies among the intended, implemented, and attained curriculum. Lynch et al. (2019) also found a positive association between student outcomes and teachers' participation in implementation meetings, which were defined as meetings in which teachers met formally or informally with other activity participants to discuss enacting intended practices. Designing structures for engagement in building shared knowledge of curriculum is explored more in Chapter 2.

Leaders can also build shared mathematics curriculum knowledge by engaging teachers and teacher teams with developing the artifacts that show connections among lessons, grade levels, and courses, thus allowing students to develop concepts in more depth. Opportunities that allow teachers to collaboratively develop the curriculum help teachers gain an understanding of the content and how the curriculum is designed. Teachers can deepen their understanding of their place in individual students' learning trajectories. They better understand how the mathematics that they place before their students affects their students' future mathematics learning experiences and how their students' prior learning experiences affect their current mathematics interactions.

Use a tool such as the Learning Trajectories online resource, available at www.mathedleadership.org/EAresources, to help teachers and leaders explore learning trajectories.

Ultimately, mathematics leaders can determine teachers' and teams' mathematics curriculum knowledge through observations and the impact on student learning (explored more in Chapter 4). *NCSM Essential Actions: Instructional Leadership in Mathematics Education* (NCSM, 2019b) describes strategies and tools you can use to collect data during classroom observations. Classroom observation tools may be downloaded from www.mathedleadership.org/EAresources.

Thus, the mathematics curriculum should be designed with the intent of supporting teachers as they teach, students as they learn, and all who support teachers and students in the process. With that in mind, let us now expand our commitment to action beyond the schoolhouse and examine ways to build a clear understanding of curriculum for all stakeholders.

Clear Understandings of Layers of Curriculum for All

A strong mathematics program makes the curriculum transparent to all stakeholders by clearly articulating how the intended curriculum is designed to meet that agreed-upon learning (what students are to know and be able to do) and assessment expectations (how students will demonstrate what they know and can do) (Barnes & Toncheff, 2016). The mathematics leader is charged with the task of ensuring a coherent and consistent implementation of the local curriculum by all teachers. This task requires a commitment to action for inclusion of all stakeholders, honoring the imperative of *commitment to others*.

Essential Action

Ensure that all stakeholders have a clear understanding of high-quality mathematics instruction and how to support it.

The articulation of the mathematics curriculum goes beyond the classroom and school building to reach all stakeholders, including families, community partners, the business community, and institutes of higher education. It is vital that resources and opportunities for stakeholders are available so that stakeholders can gain an understanding of the school's or school system's mathematics curriculum (instruction is also critical and will be explored in the next big idea, instructional leadership).

Families and community members must be included in discussions about high-quality teaching and learning, particularly when we shift how we define a student's mathematical proficiency and fluency. Kreisberg and Beyranevand (2019) note that previously, students who could compute efficiently—following a set of procedures as quickly as possible using the least amount of resources, effort, or time—were deemed to be *good at math*. Many families and community members continue to hold that view, in addition to feeling intimidated, frustrated, worried, and confused by how math is taught today. To support their understanding, Kreisberg and Beyranevand (2019) suggest helping parents understand *why* mathematics instruction has shifted (e.g., result of ongoing research and improved technology), in addition to helping families learn how they can adjust their giving of praise at home to focus more on effort and behavior and less on intelligence. Additionally, mathematics leaders can use a concrete-representational-abstract approach to show parents how students develop procedural fluency from operations that are presented with concrete manipulatives, represented with a variety of diagrams, and then generalized abstractly into procedures and algorithms.

Leaders can use guiding questions, listed in Figure 1.10, in discussions with teachers, colleagues, administrators, and families to develop resources and create opportunities that allow stakeholders, especially families, to gain an understanding of the mathematics curriculum and equip them to be advocates in ensuring that their children experience a high-quality mathematics program.

Guiding Questions	Suggested Resources or Activities
What will all students be expected to know and be able to do?	• Grade-level/course mathematics standards • Additional local or state expectations
How will students be taught?	• Balance for conceptual understanding, procedural fluency, and application • Essential features of effective mathematics instruction—the instructional strategies parents should expect their children's classroom teachers to employ (Larson & Kanold, 2016)
How will students demonstrate what the know and can do?	• Mathematical practices (CCSSO & NGA, 2010) • Mathematics proficiencies (NRC, 2001) • State and local assessment expectations
How can leaders help families support their learners?	• Provide specific suggestions on how K–5, middle school, and high school family members can help at home (Larson & Kanold, 2016) • Develop family members' general understanding K–8 through workshops—see Math and Parent Partners (MAPPS) at https://mappsua.wordpress.com • Develop elementary students' family members' understanding of the shifts in math instruction and provide them language to support growth mindset at home (Kreisberg & Beyranevand, 2019)

Figure 1.10. Questions and Resources to Support Stakeholders

Visit www.mathedleadership.org/EAresources to download a free reproducible version of this figure.

Mathematics education leaders are advocates for and expect equitable access to high-quality mathematics curriculum for all students. They must work to establish a clear shared vision of curriculum among all teachers, develop educators' mathematics curriculum knowledge, and help all stakeholders develop a clear understanding of all layers of curriculum. Additionally, leaders must advocate for and expect equitable access to high-quality mathematics instruction for all students.

Curriculum leadership is one way mathematics education leaders advocate for and expect high-quality, equitable mathematics teaching and learning. Creating a shared vision and common understanding of high-quality, relevant curriculum truly is an essential action for any mathematics education leader. The next step is to advocate for, and expect high-quality, equitable mathematics teaching and learning. Mathematics leaders bring the curriculum to life and support those they serve through instructional leadership.

Big Idea 2: Instructional Leadership for High-Quality, Equitable Mathematics Teaching and Learning

Instructional practices pertain to actions and decisions teachers or teams take to plan, implement, and reflect on lessons and student learning. "These practices include selecting and sequencing appropriate problems, tasks, activities, and questions while employing diverse approaches and representations" (NCSM, 2014b, p. 28). Instruction also includes the resources and strategies used to facilitate students' learning of the concepts and procedures as well as their ability to reason and make sense of mathematics problems.

Students can only learn the mathematics for which they have opportunities to learn; therefore, the mathematical tasks that teachers select serve as the basis of opportunity for students to learn. Teachers also set up the conditions for learning that take into account students' entry knowledge, the nature and purpose of the tasks and activities, and the kind of engagement that will be required to be successful.

Mathematics education leadership at the state, provincial, and district levels should begin with leading the appropriate stakeholders in developing a shared vision for high-quality, equitable mathematics teaching and learning and how to support it. The vision should ultimately guide the daily work of the district/province/state as well as the long-term work needed to increase student learning. What is included in the vision must be valued enough by stakeholders so that each and every stakeholder will invest in the necessary opportunities and resources to make it happen. Thus, to develop an effective vision statement, all stakeholders must participate in the process.

Effective Mathematics Teaching Practices

The vision of high-quality, equitable mathematics teaching and learning advocated in this framework is a foundational element for mathematics leadership. A strong vision requires collaboration and a common understanding of the actions that will support meeting the vision. Additionally, the second foundational element of a strong vision begins with the belief that all children can learn mathematics. Students come to the mathematics classroom having had a variety of mathematics experiences, some positive, some negative. These experiences frame the way they think about mathematics.

Essential Action

Ensure that every teacher possesses a shared understanding and vision of high-quality mathematics instruction and the actions required to meet the vision.

The goal of mathematics instruction must include ways to build students' mathematical identity. Students' mathematical identity is how and who they see themselves as in the mathematics classroom, in their relationship with the subject of mathematics and mathematical activities. Teachers and leaders both boost students' confidence, knowledge, and skills so students see themselves as doers of mathematics. When one understands mathematics, he or she sees its beauty and usefulness.

A commitment to action regarding the leader and her or his immediate sphere of influence, the imperative of *commitment to self,* focuses on developing, modeling, and leading educators in a shared understanding of high-quality mathematics instruction, particularly research-affirmed effective teaching practices.

NCTM established eight effective mathematics teaching practices (NCTM, 2014a) that are "a set of strongly recommended, research-informed actions for all teachers, coaches, and specialists in mathematics; all school and district administrators; and all educational leaders and policymakers" (p. 4).

- Establish mathematics goals to focus learning.
- Implement tasks that promote reasoning and problem solving.
- Use and connect mathematical representations.
- Facilitate meaningful mathematical discourse.
- Pose purposeful questions.
- Build procedural fluency from conceptual understanding.
- Support productive struggle in learning mathematics.
- Elicit and use evidence of student thinking.

Mathematics education leaders can use these practices to frame conversations around the vision of high-quality mathematics teaching and learning. We start by creating a culture of transparency, trust, and collaborative problem solving to build a shared vision (see Figure 1.5 for a protocol to build a shared vision) for mathematics teaching and learning. DuFour and Marzano (2011) state,

> *Leaders enhance the effectiveness of others when they provide the clarity regarding what needs to be done and ongoing support to help staff succeed.*

Remember, clear is kind! The clarity comes from creating a strong vision of what each and every student will experience and learn in mathematics and how teachers and teams collectively engage students in developing a deeper understanding for mathematics and a passion for the content.

Many of these teaching practices contradict dominant cultural beliefs about teaching and learning mathematics. To effect change in teacher behaviors in these areas may require bold instructional leadership. NCSM (2019a, p. 53) identifies four actions that leaders can take to support the implementation of these practices and other forms of high-quality instruction.

- Ensure every teacher and leader is able to define high-quality instruction.
- Teacher teams use standards and data to ensure consistent learning expectations both horizontally (across grade level) and vertically (between grade levels).
- Ensure teacher teams identify mathematical tasks that promote reasoning and problem solving and vary to address multiple levels of cognitive demand and that teacher teams have the knowledge and ability to do so.
- Collaborative planning conversations are focused on instruction and student learning.

Mathematics education leaders can help teachers cultivate these teaching practices using a set of continua like the ones shown in Figure 1.11. Teachers can use the descriptors along the continua to reflect on their current status with regard to each effective mathematics teaching practice.

Mathematics Teaching Practices Continuum

What do they look like in practice?

Teacher Centered - - - - - ➤	Making Progress - - - - - - - - - - ➤	Student Centered
ESTABLISH MATHEMATICS GOALS TO FOCUS LEARNING.		
Students are specifically reminded of the mathematics they did previously and what they will do next. Teacher shows general rule or talks about the point of the lesson.	Mathematics goals are clearly or become clearly related to the mathematics students have been learning and the overall goals of the unit, grade level, or course and possibly to a familiar situation or future application. Lesson includes an opportunity for students to generalize, explain, and justify solutions. Mathematics goal drives the student-led summary.	Mathematics goals of the lesson are clear and connected to the learning progression or become clear and connected to the learning progression. Students can explain why they are doing a mathematical task, how it is related to other areas of math, and possibly some important uses. Lesson engages all students in generalizing, explaining, and justifying solutions. Mathematics goal drives the student-led summary and the connections. Students can explain why they are doing the task, how it is related to other areas of math, and possibly some important uses.

Figure 1.11. Sample of One MTP Continuum
Source: Adapted from NCTM (2014a, 2017); McGatha, Bay-Williams, Kobett, and Wray (2018); and Delaware Math Coalition (2013).

Teachers and teams can reference the continuum to self-assess their current strengths and areas for growth. When working with other leaders who support mathematics teachers, the continuum paints a picture of a vision for high-quality mathematics instruction.

Visit www.mathedleadership.org/EAresources to download a free reproducible version of all eight mathematics teaching practices continua.

As teachers are implementing the effective mathematics teaching practices, they should do so within a culture of access and equity. In its position statement about *Access and Equity in Mathematics Education,* the National Council of Teachers of Mathematics (NCTM) states that creating, supporting, and sustaining a culture of access and equity require being responsive to students' backgrounds, experiences, cultural perspectives, traditions, and knowledge when designing and implementing a mathematics program (NCTM, 2014b).

Structures for designing and implementing a mathematics program will be explored more in Chapter 2. However, a commitment to action for engaging colleagues in culturally relevant teaching supports a culture of access and equity. Effective leaders guide teachers down this pathway.

Culturally Relevant Teaching

If we want to reach students who have traditionally been left behind in the mathematics pipeline, then we must make culturally relevant teaching a focus in mathematics classrooms. Teachers must seek opportunities to learn more about the communities from whence their students come. A leader's commitment to action supporting colleagues outside their normal sphere of influence, the imperative of *commitment to colleagues,* focuses on the design and implementation of meaningful student learning experiences. With an eye on equity, culturally relevant teaching provides a useful lens through which to view what makes a student's learning experience meaningful. Consider one leader's journey to uncover her understandings of cultural relevance in relation to mathematics instruction.

Essential Action

Ensure that every teacher possesses the skills and knowledge necessary to design and implement meaningful learning experiences that lead to student understanding of mathematics.

"Culturally relevant pedagogy is a pedagogy that empowers students intellectually, socially, emotionally, and politically by using cultural referents to impart knowledge, skills, and attitudes" (Ladson-Billings, 1994). As my professor posted that quote on the board in our graduate class, I came to the abrupt realization that I had never heard about how pedagogy could empower students. I knew in that moment that it was imperative that I learn more about culturally relevant practices and how they would help both impact the teachers I coached and their students. What I had not anticipated in this moment was an intersection where my childhood and my teaching career would collide and create an unforeseen moment of healing. As I worked my way through school from kindergarten through college, I had never seen glimpses of myself or my culture represented in the curriculum, lessons, materials, or artifacts in my classrooms. I could see representations of my classmates and their families in stories, movies, worksheets, poems, and math problems but I never saw myself. At that time, I did not understand why it was important or what I was seeking, but as I reflect, I now realize I was seeking a sense of belonging and to be seen.

As I learned more about culturally relevant practices, I decided that the healing process would mean being bold and sharing with teachers and students a part of myself. Stepping out of fear and into courage meant sharing my journey and culture. I began by asking two teachers I had built a strong relationship to teach a mathematics lesson to their classes. I created and launched the lesson by using the window and mirror analogy from the work of Rudine Sims Bishop (1990), later adapted to mathematics by Rochelle Gutiérrez (2012). I explained to students that I was going to share a story problem with them and the mirror was a view of myself and my culture, but it would be a window for them into my life and culture. The story problem was about a birthday cake that I would have on my birthday called "Bolo de Bolacha" that my mom would make. It was made out of layered cookies and frosting. The story problem was aligned to a third-grade multiplication standard.

Mrs. Rivera was making a bolo de bolacha for her mom's birthday. She used 7 cookies on the bottom layer of the cake. The cake was 10 layers high. How many cookies did she use to make her cake (bolo)?

After sharing the story problem with them, I challenged the students to write a story problem that was a mirror of their culture and a window for all of us in the class. As a class, we had to unpack what the word culture meant. Students discovered that it went beyond race or ethnicity because culture includes traditions, hobbies, art, music, customs, family interests, their neighborhood, environment, or religion. In my experience, it is difficult to get students to write story problems, but in this case, I could not get them to stop. They were elated to research ideas online, look at problem types, and then share their stories through problems they were sharing with their classmates. As I stood in the back and watched students say, "Here is a mirror of me, let this be a window for you of who I am," tears streamed down my face because in that moment, the little girl that was never seen in her mathematics courses was now creating a window for all the students in the room for their culture to be seen through mathematics story problems. In the end, I collected all their stories and put them in a book that I keep to remind me that being bold made a difference.

Now that I am in a leadership position, I have been able to take what I have learned about culturally relevant pedagogy to help guide an algebra textbook committee to look at mathematics textbooks with a different lens. I posed the question, "Do the problems in the textbook reflect our students and their culture?" The prompt opened up the discussion about culturally relevant practices and how to view rigorous math tasks that provide a window or mirror for our students. As a result, the team selected a text that was aligned to rigorous standards but also reflected and opened up windows for our student population. In another circumstance, I led a curriculum writing team and was able to incorporate a culturally relevant rubric to rate how the curriculum was culturally relevant, which led the team to make adjustments and reframe their thinking. Lastly, I continue to share my story with the teachers and leaders I serve locally and at national conferences in hopes that someone else in a mathematics leadership position will learn more about culturally relevant practices and then take the leap of faith to learn more and be a BOLD leader by incorporating culturally relevant practices as they make leadership decisions to ensure equity and access for each and every learner.

Georgina Rivera, NCSM Professional Learning Director (2019–2021)

For this teacher turned leader, it was not that she did not want to teach using a culturally relevant approach; it was that she did not know it even existed. Coming from a minority background in her K–12 schooling, she knew how it felt to not see herself in the mathematics curriculum. For her, the change was not difficult because she had felt the ping of not belonging. For some teachers, it may be difficult to change the way they teach; therefore, just as students need opportunities to learn, so do teachers. If we expect substantial changes are needed to create more equitable classrooms, then we must provide the necessary learning for teachers to teach meaningful and relevant mathematics, just as this leader experienced an opportunity to change her thinking about mathematics teaching.

You can download a PDF version of Rivera's story at www.mathedleadership.org/EAresources. Her story may be used as a case study in a professional learning community text-based discussion. Use a discussion protocol such as the one shown in Figure 1.12 to guide the discussion.

Text-Based Discussion Protocol

Preparation: Timing is important as it keeps the focus on building on each other's thinking rather than engaging in dialogue. Participants should work in small groups of four.

Protocol:

1. In each group, select a timekeeper to keep pace.
2. Each participant should select a quote or brief passage from the article.
3. One member volunteers to go first. This presenter identifies the quote or passage and then reads it aloud without explanation.
4. Each of the other three members gets 1 minute to respond to the passage, which could be used to say what the passage makes them think about, what questions or comments it raises, and so on.
5. The first participant takes 3 minutes to state why she or he chose that passage and then responds or builds on what her or his colleagues have stated.
6. Repeat the pattern until all participants have been a presenter.
7. Come back to the whole group. Have each small group share a key idea or thinking that emerged from the discussions.
8. Debrief the process. How did this structure help participants explore ideas in the text and help participants connect it to their own thinking?

Figure 1.12. Text-Based Discussion Protocol
Source: Adapted from National School Reform Faculty.

Visit www.mathedleadership.org/EAresources to download a free reproducible version of this figure.

To connect mathematics to students' experiences in productive ways, teachers need to develop ongoing practices around learning about their students—their students' interests, everyday activities, heritage, home languages, and more (e.g., Emdin, 2016; Gay & Kirkland, 2003). Teachers also need the mathematical and pedagogical comfort and confidence, along with support from administration, to adapt common textbook problems to make them more applicable. For example, in a sixth-grade urban classroom in Worcester, Massachusetts, a teacher took a lesson from her textbook that focused on sorting paper clips using ratio language and changed the sort to popular sneaker brands that her kids are constantly discussing. In doing so, she had 100% student engagement because the kids found the mathematics relevant to them—and were excited to share their noticings, such as "there are 2 pairs of Nikes for every 4 pairs of Converse sneakers." She felt confident adjusting the curriculum in this case because her administrators have set clear expectations around maintaining the intended mathematical rigor, while adjusting content to be more culturally relevant.

Leaders and teachers who choose to commit themselves to providing culturally relevant mathematics experiences are taking actions that support enacting meaningful and relevant mathematics. How do we know when the math is meaningful and relevant to students? Matthews, Jones, and Parker (2013) put forth the *Culturally Relevant Cognitively Demanding Mathematics Task Framework* (CRCD rubric) to assist teachers in creating tasks that are relevant to their students. The framework can serve as a tool to guide teachers in the selection of tasks that are cognitively demanding (Smith & Stein, 1998) coupled with appropriate social and learning conditions extracted from the existing literature on culturally responsive teaching (Gay, 2000) and culturally relevant pedagogy (Ladson-Billings, 1994).

Since the first publishing of the framework in 2013, it has been refined and is shown in Figure 1.13. The developers of the CRCD rubric use ***emerging, developing,*** and ***exemplary*** to define the products of educators' underlying purpose of mathematics tasks.

Task Level	Description of Mathematical Task Level
Emerging	**Requires considerable cognitive effort in mathematics** • Task is mathematically rich and cognitively demanding. • The task requires considerable efforts using multiple representations and strategies to develop deep understanding of mathematics. • Task content draws from connections to other relevant subjects, disciplines, and concepts.
Developing	**Requires considerable cognitive effort AND is embedded in cultural/self/ community inquiry and activity** • The task is centered in real-world situations requiring students to inquire deeply about themselves, their communities, and the world about them. • Requires students to draw from, use, and embrace community and cultural knowledge directly in developing strategy and solution process. • Task content seeks to add to this knowledge through mathematical activity.
Exemplary	**Requires considerable cognitive embedded in cultural inquiry and activity AND targets cultural/self/community empowerment and social justice** • The task requires students to examine structure and assumptions of self, community, the world, and its relations in considerations of solutions and strategy limits. • Task requires students to examine conditions of opportunity, justice, suffering, and inequity that arise in their communities, schools, and the world around them. • Task utilizes mathematical sense-making and the solution processes to help students to develop informed perspectives and take action on real-world issues.

Figure 1.13. The Revised CRCD Mathematics Task Framework
Source: Adapted from Matthews, Jones, and Parker (2013).

Visit www.mathedleadership.org/EAresources to download a free reproducible version of this figure.

In essence, educators committed to culturally responsive teaching in mathematics tasks will create, encourage, and use tasks that require considerable cognitive effort in mathematics; are embedded in meaningful cultural, self, and community inquiry; and deliberately target cultural, self, and community empowerment and social justice.

As leaders, using the CRCD rubric provides clarity for teachers and teams to assess the richness of the mathematics as well as the depth of cultural and community knowledge students are required to access directly through the task. The rubric can also be used to assess the possible impact on students' lives directly. Lastly, the rubric aligns the purposes of the task to social justice and empowerment.

As we have seen in practice, a critical consciousness remains challenging for educators (Jones, 2018). That is, educators remain reluctant to critique the aims of mathematics and repurpose these aims for pursuits of justice. The rubric poses a progression that teachers move from emerging, or less critical, to exemplary, or more critical, orientations.

In order to move forward in your work of creating equitable classrooms through the use of culturally relevant mathematics tasks, mathematics leaders must reflect on your current level and use of CRCD mathematics tasks. Use the CRCD Mathematics Reflection Tool in Figure 1.14 to gather evidence from the current level of CRCD Mathematics at the classroom, school, or district level.

Current Level of CRCD Mathematics	**Evidence of Current Level** ***List the Evidence in the Classroom and/or School.***
Emerging—Products of educators who believe that the basis of responsiveness lay in crafting tasks that challenge all students and that all students are capable of doing and succeeding in mathematics. These tasks reflect high-quality features of problem solving and cognitive demand.	
Developing—Products and activities where educators use broadened definitions of mathematics as cultural activity and design task prompts, features and interactions that require students to also do cultural, community, and self-exploration and affirmation. While these tasks represent some progress in design, they are not yet culturally responsive because they do not "problematize" the status quo or offer critical inquiry into race, culture, inequality, and social justice.	
Exemplary—Products and activities where educators used broadened definitions of mathematics, cultural inquiry, and social justice for personal and community pursuits. These tasks feature student action, critical investigation, community affirmation, and empowerment notions central to a culturally relevant/responsive pedagogy.	

Figure 1.14. Reflection on Culturally Relevant Mathematical Tasks

Visit www.mathedleadership.org/EAresources to download a free reproducible version of this figure.

Instructional leadership with a focus on culturally relevant pedagogy (NCSM & TODOS, 2016) and culturally relevant teaching practices helps ensure that every teacher possesses the skills and knowledge necessary to design and implement meaningful learning experiences that lead to students' understanding of mathematics. Leaders can help ensure that every teacher possesses a shared vision of high-quality mathematics teaching by helping them better understand effective teaching practices. Leaders can ensure that every teacher and team possess the skills for meaningful and relevant instruction by focusing on culturally relevant teaching practices.

Clear Understandings of High-Quality Instruction

Essential Action

Ensure that all stakeholders have a clear understanding of high-quality mathematics instruction and how to support it.

A commitment to action supporting all stakeholders in the greater school community, the imperative of *commitment to others,* focuses on ensuring that all stakeholders, within and outside of the school, have a clear understanding of what high-quality mathematics instruction looks like and how they can support it. This challenge can create tension between schools and constituents such as parents if it is not well attended to since research-informed effective instructional strategies do not look like the ones that many adults experienced in their own mathematics instruction.

Students come from families and communities that have accumulated social and cultural knowledge. Moll, Amanti, Neff, and Gonzalez (1992) use "the term 'funds of knowledge' to refer to these historically accumulated and culturally developed bodies of knowledge and skills essential for household or individual functioning and well-being" (p. 133). When teachers and schools make the effort to learn about their students and families, teachers can use those "funds of knowledge" to engage students in culturally relevant mathematics lessons, exemplifying the foundational element of *relationships.* This element reminds us that to implement and achieve a vision for mathematics teaching and learning for all, we must work collectively and build working relationships among all stakeholders. Using culturally relevant teaching practices strengthens relationships among students, teachers and students, teachers and parents, and schools and community. When these relationships are strengthened, students win.

There are specific actions that leaders, based on their leadership role, can take to work toward this goal. Some of those actions are described in Figure 1.15.

Using culturally relevant teaching practices strengthens relationships among students, teachers and students, teachers and parents, and schools and community. When these relationships are strengthened, students win.

Mathematics Leadership Role	Possible Actions for the Specific Role
Teacher Leader	• Take the time to understand how negative stereotypes impact their students and actively work to erase the effects of those negative stereotypes. • Make better connections to students' prior knowledge inside and outside of school. • Ensure that equity is central to your work, not an add-on to your work. • Commit to providing educational products that support student engagement of the Standards for Mathematical Practice or other mathematical processes found in provincial or state standards. • Use culturally relevant curriculum materials and resources that highlight the contributions of diverse people to mathematics and STEM. • Release control in the classroom (i.e., empower students to take ownership of their own learning by focusing on sense-making). • Position all students as mathematically capable.
Coach	• Help teachers connect to students' funds of knowledge. • Make better connections to students' prior knowledge inside and outside of school.
Site Leader	• Create structures for discussing and implementing an equity framework in a systematic way. • Help teachers to develop a critical consciousness regarding your school/district and use that as a basis for decisions to eliminate inequities.
District Leader	• Support teachers and schools in their efforts to learn about and understand their students, families, and communities in order to use this information to improve instruction. • Take the time to understand how negative stereotypes impact their students and actively work to erase the effects of those negative stereotypes.
State/Provincial Leader	• Create emotionally safe spaces for meaningful dialogue for all stakeholders, including business and parent groups. • Initiate critical conversations with policy influencers regarding inequitable outcomes for students who have traditionally been left out of high-quality and meaningful mathematics.

Figure 1.15. Leadership Actions Supporting Clear Understandings of Instruction Among All Stakeholders

Use a tool like the one shown in Figure 1.16 to reflect on how you currently use leadership actions for your role to support clear understandings of instruction among all stakeholders. Identify at least two key actions for your role and consider how you are doing those actions now, what else you could be doing to support that action, and what resources you need to extend your work with that action.

Leadership Role:			
Key Action:	How am I doing this now?	What else could I be doing?	What resources do I need to do it?
Key Action:	How am I doing this now?	What else could I be doing?	What resources do I need to do it?
Key Action:	How am I doing this now?	What else could I be doing?	What resources do I need to do it?

Figure 1.16. Reflection Tool for Leadership Actions

Visit www.mathedleadership.org/EAresources to download a free reproducible version of this figure.

When teachers have equity at the center of their work, there is more opportunity to create lessons that are equitable and that empower students. Students must be able to construct meaning through the work they do in the classrooms as it relates to their immediate contexts. Teachers must learn about the students and families they serve in authentic ways. School leaders and teachers must reach out to the allies in their communities who care about improving the achievement of all students. They should partner with community groups who have a track record of positively impacting students and who stay the course and finish the commitments they make when working with schools.

Armed with knowledge about students gleaned from working with all stakeholders, teachers must then plan how they would facilitate a culturally relevant lesson that are anchored in effective mathematics teaching practices (NCTM, 2014a). The mathematics task is at the heart of "how students make sense of [the] mathematics" (Harbin Miles et al., 2019, p. 155). These lessons should

- Provide students opportunities to explain and justify their thinking
- Include a high level of cognitive demand and a culturally relevant context
- Be supported with appropriate scaffolding questions that maintain the level of cognitive demand

It is especially important that teachers anticipate student misconceptions and plan to address them during the instructional activities. The questions that teachers ask will help to facilitate effective mathematical discourse. "When you plan questions to ask during a lesson, you also increase access and engagement for all students because you are mindful about whom you will be calling on" (Harbin Miles et al., 2019, p. 155). This will be explored more in Chapter 2.

As a leader, you are challenged with uncovering and addressing inequities that exist in mathematics teaching and learning (NCSM, 2019b). In *NCSM Essential Actions: Instructional Leadership in Mathematics Education,* you are taken through the process of building a shared leadership model by developing a schoolwide mathematics school improvement team (MSIT). This team is a great way to assess your needs as a school. This process can also be done at the district level and should involve all the appropriate stakeholders.

Summary of Essential Actions

Effective mathematics education leaders **advocate** for and expect high-quality and equitable mathematics teaching and learning for all students. Not some, but all. As you have seen in this chapter, curriculum leadership and instructional leadership are two key ways that leaders can provide this advocacy and expectation. Curriculum leadership that advocates for and expects high-quality and equitable mathematics teaching and learning includes helping teachers develop a clear shared vision for what high-quality curriculum looks like. Curriculum leaders cultivate teachers' and other educators' mathematics curriculum knowledge. Curriculum leaders work with all stakeholders to develop a clear, shared understanding of what high-quality curriculum looks like so that all students graduate high school ready for careers, college, or military success.

As well, instructional leadership is key. Curriculum defines *what* students should learn and instruction defines *how* teachers should teach it. Instructional leadership includes providing teachers ongoing opportunities to study and sharpen effective mathematics teaching practices. Instructional leaders cultivate their teachers' culturally relevant teaching practices. Effective instructional leaders work with all stakeholders, including parents and community members, to develop a clear, shared understanding of the research-affirmed instructional strategies that will help students learn the mathematics necessary for their future success.

Curriculum defines *what* students should learn and instruction defines *how* teachers should teach it.

Taking Action

Figure 1.17 is a reflection on the three imperatives aligned to the first guiding principle: **Advocate** high-quality, equitable mathematics teaching and learning for every student. Complete the tool to reflect on where you are now in your own leadership role with regard to each action step. Then, identify potential next steps that you can take to work toward each action.

Action Step	Reflection Questions	Next Steps
Imperative 1: *Ensure that every teacher possesses a shared understanding and vision of high-quality mathematics instruction and the actions required to meet the vision.*		
Help teachers deepen their knowledge of the different layers of curriculum.	How well do our intended curriculum and implemented curriculum align? What additional resources do we need in order to increase that alignment?	
Develop teacher knowledge of effective mathematics teaching practices.	What do each of the effective mathematics teaching practices look like in the classroom?	
Imperative 2: *Ensure that every teacher possesses the skills and knowledge necessary to design and implement meaningful learning experiences that lead to student understanding of mathematics.*		
Help teachers develop mathematics curriculum knowledge.	What opportunities do teachers have to study their grade/course-level curriculum in vertical teams?	
Develop teacher knowledge of culturally relevant teaching.	What opportunities do we provide for teachers to use student interests as contexts for the mathematics?	
Imperative 3: *Ensure that all stakeholders have a clear understanding of high-quality mathematics instruction and how to support it.*		
Collaborate with all stakeholders in order to build a shared vision of high-quality mathematics curriculum that is aligned to state/provincial expectations and community priorities.	How well do all stakeholders understand the provincial/state curriculum standards? How do the provincial/state curriculum standards support community needs and interests?	
Collaborate with all stakeholders in order to articulate high-quality mathematics instruction, how it benefits students, and how stakeholders can best support the instruction.	What opportunities can we provide for stakeholders to witness high-quality, equitable mathematics instruction in action? How are stakeholders invited to participate in the development of curriculum or instruction resources?	

Figure 1.17. Self-Reflection on Guiding Principle 1

Visit www.mathedleadership.org/EAresources to download a free reproducible version of this figure.

In this chapter, you learned more about the first guiding principle: **Advocate** and expect high-quality, equitable mathematics teaching and learning for all students, including key elements of curriculum leadership and instructional leadership. In the next chapter, you will dig deeper into the second guiding principle: **Design** and implement structures that support high-quality mathematics teaching and learning for every teacher.

CHAPTER 2

Guiding Principle 2: Design

"Every teacher works interdependently in a collaborative learning community to erase inequities in student learning."

—PRIME Equity Indicator 3 (p. 18)

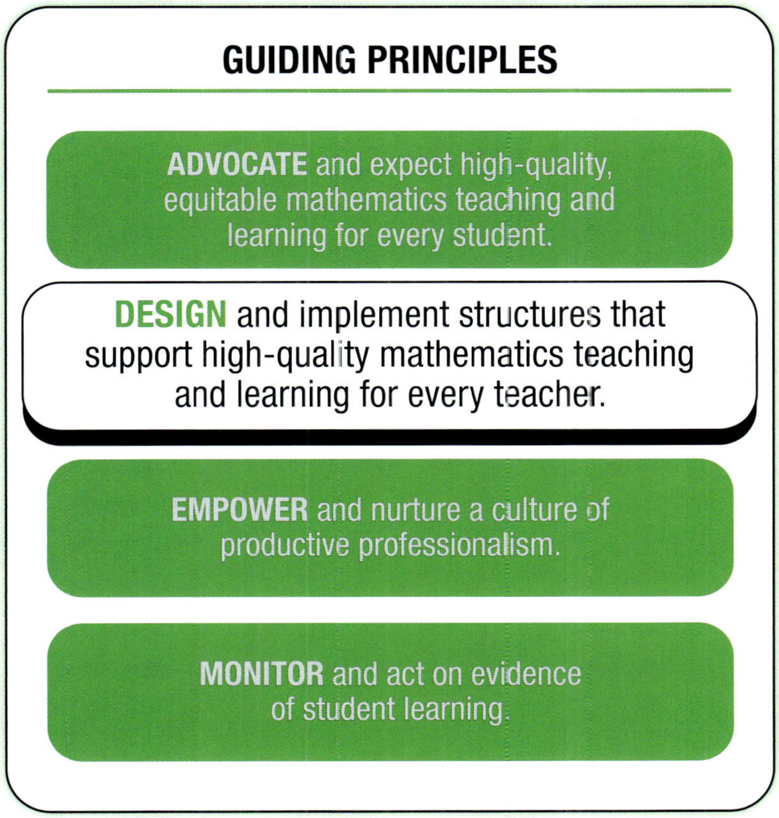

Guiding Principle 2: DESIGN and implement structures that support high-quality mathematics teaching and learning for every teacher.

As leaders **advocate** for students to receive high-quality mathematics instruction, they need to ensure that structures supporting the work of teachers exist so that desired outcomes are likely to be met. The second principle, **design,** provides clarity for the work of mathematics leaders and requires that each type of leader (classroom level, team level, school level, district level, state/province level) design and implement structures that support high-quality mathematics teaching and learning for every teacher. Consider the scenarios in Figure 2.1.

Scenario 1: A district mathematics leader examines the catalog of approved courses to identify course names, course prerequisites, and other practices or processes that serve to limit student access to rigorous mathematics. The leader identifies a Grade 9 course titled "Introduction to Algebra" as a course that delays access to high school–level mathematics and notes, through a close examination of course enrollment, that predominantly African American, Hispanic, and students receiving special education services are accessing the course. The leader commits to a plan for eliminating this course within the next 2 years.
Scenario 2: A middle school mathematics coach collaboratively designs and leads professional learning for her team. The professional learning is focused on the selection and effective classroom implementation of mathematical tasks and the facilitation of effective mathematical discourse. The mathematics team investigates research-affirmed instructional strategies and integrates those ideas into lessons. Teachers informally observe their peers, providing formative feedback focused on the professional learning target outcome.

Figure 2.1. Design and Implement Structures Scenarios

In each scenario above, evidence of an impact on student learning was collected and acted upon. In the first scenario, a district-level leader noticed a course pathway that impeded students' access to rigorous mathematics, particularly for students who have been systematically excluded from high-quality mathematics teaching and learning. The leader then collected data and committed to a plan to eliminate this barrier to equitable access.

In the second scenario, teachers engaged in peer observation, collecting observational data to provide one another with formative feedback to improve their own practice and improve student learning. They could do so because their campus administration created structures that supported their collaborative learning, including a coach and opportunities for teacher teams to meet, plan, observe instruction, collect data, and take action. Whether at the classroom level, team level, or district program level, it is important to consider how existing structures and practices either enhance or limit a student's mathematical journey.

This chapter focuses on the essential actions and imperatives aligned to the Design principle. The imperatives and their associated essential actions are described in Figure 2.2.

	Imperative 1 ***Commitment to Self***	**Imperative 2** ***Commitment to Colleagues***	**Imperative 3** ***Commitment to Others***
DESIGN and implement structures that support high-quality mathematics teaching and learning for every teacher.	**Essential Action** *Ensure mathematics learning for all students through organizational structures, time and resource allocation, and systemic supports that are aligned, intentional and equitable.*	**Essential Action** *Ensure systems of continual collaborative, job-embedded professional learning to build teacher and leader capacity and increase efficacy.*	**Essential Action** *Ensure sustainability through engaging all stakeholders in systemic, long-range strategic planning for all teaching and learning improvement initiatives.*

Figure 2.2. Design Guiding Principle From the Framework for Leadership in Mathematics Education

Within the design guiding principle, the commitments to action among the different imperatives emerge as follows.

- Actions related to the first imperative, *commitment to self,* focus on how a leader works with educators in her or his sphere of influence to self-examine current beliefs and actions regarding access and equity, support curriculum and intensification programs as a response to student learning, and design and implement effective structures for professional learning.

- Actions related to the second imperative, *commitment to colleagues,* focus on how a leader works with educators in adjacent spheres of influence, such as other teacher teams, other departments, on a campus, or within the district, to create structures that can influence teachers' beliefs and actions about equity and access, build teachers' collective efficacy, and prioritize collaborative professional learning.
- Actions related to the third imperative, *commitment to others,* focus on how a leader works outside of his or her immediate learning community in order to effect systemic changes, such as detracking, that supporting equity and access, engage stakeholders, and develop a coherent and focused professional learning plan.

The **Design** imperatives show the importance of developing structures that ensures that students are receiving high-quality instruction from well-trained, highly skilled teachers whose work is aligned to a clearly articulated strategic plan. This chapter is designed to help bring these imperative to action by focusing on:

- Big Idea 1: Designing and implanting structures to support access and equity
- Big Idea 2: Designing and implementing structures to support mathematics teaching and learning
- Big Idea 3: Designing and implementing structures to support professional learning

As a mathematics leader, what is your role in designing and implementing structures that support high-quality mathematics teaching and learning for every teacher? Use the guiding questions in Figure 2.3 to reflect, based on your leadership role, on how you as a bold leader can support the three essential actions within the **Design** guiding principle.

Guiding Principle 2: DESIGN and implement structures that support high-quality mathematics teaching and learning for every teacher.

Essential Action	Guiding Questions
Ensure mathematics learning for all students through organizational structures, time and resource allocation, and systemic supports that are aligned, intentional and equitable.	In my leadership role: • What structures do I work within that support access and equity for all students? • How do I support first-time instruction and necessary intensification and intervention? • How do we allocate resources to support collaborative professional learning?
Ensure systems of continual collaborative, job-embedded professional learning to build teacher and leader capacity and increase efficacy.	In my leadership role: • How do I support professional learning about culturally responsive teaching? • What structures do we have in place that support collaborative planning and lesson development? • What structures do we have in place that support collaborative, job-embedded professional learning?
Ensure sustainability through engaging all stakeholders in systemic, long-range strategic planning for all teaching and learning improvement initiatives.	In my leadership role: • What structures can we use as we work toward detracking teachers and students? • How does my work support the campus or district mathematics improvement plan? • What is the long-range plan for professional learning in my context?

Figure 2.3. Leadership Roles With the Design Guiding Principle

Visit http://www.mathedleadership.org/EAresources for a role-specific reflection tool with questions for each mathematics leadership role (teacher leader, coach, site-based leader, district leader, and state/provincial leader).

As we continue with the **Design** guiding principle, consider your reflection in Figure 2.3 to plan next steps.

Big Idea 1: Designing and Implementing Structures to Support Access and Equity

For decades, there has existed a shared goal of ensuring that each and every student experiences success in mathematics. Meeting that goal has been a constant challenge. In order to overcome this challenge, mathematics leaders must confront the systems, structures, and biases that are pervasive not only in their school or district but also in society as a whole. It is the responsibility of mathematics leaders to develop and support structures that promote access and equity for all students.

Access, in this chapter, refers not only to the opportunities afforded all students but also the structures designed to ensure that those opportunities are likely to result in student learning success. The term *equity* does not mean *equal.* Supporting equitable mathematics learning means that mathematics leaders are working to remove barriers to access, to provide the customized support needed by each student, and to ensure that teachers have the tools, knowledge, and resources to help each and every child learn. The *access and equity* challenge experienced over the past few decades persists, in part, because existing structures, belief systems, and policies serve to exclude some students.

A goal for all mathematics leaders is to look inward in order to understand the broader systems of inequity that exist in society as well as the systems of inequity that permeate a school district. We are charged to leverage the understanding of the inequities in order to provide access and opportunities to students representing traditionally underserved or underrepresented populations. Further, mathematics leaders examine their own processes and practices in an effort to identify bias and then work to mitigate or eliminate systems that perpetuate exclusionary practices.

Equitable Learning Opportunities Begin With Self-Examination

Essential Action

Ensure mathematics learning for all students through organizational structures, time and resource allocation, and systemic supports that are aligned, intentional, and equitable.

NCSM's 2019 monograph, *Acknowledging Our Role in the Education Debt,* encourages mathematics leaders to build self-knowledge about the broader system of inequities in society that impact children of color and poor children's lack of opportunities in learning high-quality mathematics. When considering structures that support access and equity, building self-knowledge is an important action for a *commitment to self* in your leadership role (NCSM, 2019c).

The Annie E. Casey Foundation (2017) identifies 12 key milestones for future success in life, including attaining eighth- grade mathematics proficiency on time as measured by the National Assessment of Educational Progress (NAEP) standards and graduation from high school in 4 years. Although many people overcome setbacks in school learning and in reaching life's goals, it is particularly harder for those impacted by racism and other systemic injustices.

For example, Turner (2018) focuses on five domains that influence a person's life outcomes: housing, health, criminal justice, employment and entrepreneurship, and education, with education having the most influence. For each domain, there is historical evidence of decades of discriminatory practices and unfair policies that continue to plague African American, Hispanic, Native American, and poor families on a variety of levels.

The Importance of Self-Examination

Effective leaders of mathematics understand this and know that before deep work in equity can begin, it is necessary to acknowledge injustices and unfair policies. Structured time is needed for educators to look within and examine personal and collective roles in maintaining such inequitable systems.

Figure 2.4 is a reflection tool designed to help mathematics leaders look inward to reflect on current practices. The left-hand column lists a series of leadership actions taken from the joint NCSM and TODOS (2016) position paper *Mathematics Education Through the Lens of Social Justice: Acknowledgment, Actions, and Accountability.* Read each item from this list and reflect on your current state. In the right-hand column, develop a list of actions you would take to advance access and equity work in your school, district, state, or province.

Intended Leadership Actions	Personal Reflections and Next Steps
Interrogate individual and societal beliefs underlying the deficit views about mathematics learning and children with specific attention to race/ethnicity, class, gender, culture, and language.	
Refrain from using deficit discourse in professional learning communities and instructional decision making (e.g., placement decision, course offerings, intervention strategies).	
Eliminate tracking systems that sort children based on perceived ability and demographic profile.	
Show evidence that course-taking patterns are changing, remedial/intervention courses are reduced, and advanced mathematics offerings are more robust and plentiful.	
Increase recruitment and retention of mathematics teachers and leaders from historically marginalized groups.	
Create fair and holistic assessment systems for students and teachers of mathematics that provide productive and timely information on learning and are free from high-stakes pressure, static labeling of students and schools, and arbitrary sanctions.	
Require professional development opportunities that focus on social, cultural, linguistic, contextual, and cognitive facets of mathematics and mathematics learning.	
Create a mathematics vision with accountability mechanisms for the classroom, school, and district that uplifts students to learn rigorous and relevant mathematics.	

Figure 2.4. Reflection on Leading Through the Lens of Social Justice

Visit www.mathedleadership.org/EAresources to download a free reproducible version of this figure.

Deficit Terminology

Looking within is a self and collective learning process. The reflective process sets the stage for clarity in developing structures to address deficit terminology, effective and equitable teaching practices, professional development, and detracking of students and teachers.

Think about the last time you were in a team meeting or staff meeting. What language was used to describe the students who were struggling with the latest content?

Effective leaders of mathematics challenge the use of deficit terminology when discussing children and parents of color and poor families. Mathematics leaders encourage personal examination of beliefs that lead to deficit views about learning based on race/ethnicity, class, gender, culture, language, or disability that "constrain discourse and understanding" (Confrey, 2010). Language that refers to students as "the slow kids" and "the SPED kids" is oppressive and sends a message that some children do not measure up to a perceived dominant standard of social and academic norms. Instead, effective mathematics leaders encourage the use of strength and abundance messages that emphasize that all children are smart and add culture and curiosity to their learning (Jilk, 2016). Deficit terminology leads to deficit actions, making it more difficult to develop meaningful solutions for underachieving students.

Toncheff (2018) provides specific ways to eliminate some deficit terminology that is commonly heard in team discussions. Figure 2.5 provides different ways of phrasing concerns that are not rooted in deficit terminology.

Instead of thinking or stating:	Discussion can be centered on...
Those students can't...	This group of students (specify how many and state names) is struggling with engagement in the lesson. What might we do differently to support their continued learning?
They do not want to learn.	This group of students (specify how many and state names) is shutting down early in the lesson. Which strategies have you found successful to encourage productive perseverance? Which strategies are developing a growth mindset in students and helping them embrace mistakes as on opportunity to continue to learn?
They are not motivated, or they do not even try.	Student C and Student D have been scoring low on assessments and are choosing unproductive behaviors during class. What can we do differently to keep the students engaged? How can we make the mathematics in the upcoming lesson come to life, so students can see the power of the mathematics?
They don't do anything.	Student A and Student B have not submitted their independent practice for the last week. What required intervention can we provide to ensure that the students are given additional time and support to identify if the students are struggling on the essential standards or if they are failing to complete the assignments for other reasons?
These students are low.	This group of students (specify how many and state names) is struggling with the two of the learning targets this week. What is the prerequisite knowledge needed to access the content in the lessons next week? How can we plan to tap into the prior knowledge needed so we can target specific skills to support learning during intervention or small group time?

Figure 2.5. Rephrasing Deficit Terminology

Student tracking systems and policies that sort children by perceived ability are examples of deficit actions that occur as a result of beliefs that some children do not have the same capacity to learn mathematics as others. The concept of tracking is revisited later in this chapter.

Additionally, structures are needed to establish equitable grading practices. Traditional grading practices use a mixture of quantitative and qualitative measures to determine grades (Feldman, 2019). Aware and unaware teacher bias, stereotypes, and attitudes often influence grades by including unfair practices such as deducting points for student behaviors, participation grades, factoring the amount of homework submission and performance, and assumptions made about academic ability based on race and social class. Such practices are viewed as punishment and often imposed on students whose culture is different from the teacher. Additionally, mathematics leaders encourage grading systems that communicate what a student knows and is able to do. Instead, mathematics leaders work to develop grading practices that focus more on student end-of-unit performance and growth rather averaging performance over time. End-of-unit performance more accurately represents content mastery and levels the playing field for all students.

The goal of providing equitable access to a rigorous mathematics course for all students may have presented challenges to educators over the past few decades, but a commitment to its attainment begins with self-examination and a close examination of existing processes, procedures, and policies. In the next section, leaders examine what it means to lead others toward equitable mathematics programming.

Learning and Leading With Equity

After you have identified potential actions or structures to adjust to ensure equitable instruction, you will need to identify professional learning opportunities to close the gap between your vision and current reality. A *commitment* to action for *colleagues* that you can take is to lead professional learning that is equity focused and designed to address the influence of implicit bias in our systems designs as well as the personal, often unconscious, bias that influences each educator within the system.

NCSM and TODOS (2016) assert that

> *an increased understanding of mathematical content knowledge is vital, yet insufficient for 21st century mathematics teaching. Equally important, teachers and leaders need on-going research-based professional learning that focuses on the sociopolitical turn of mathematics education (Gutiérrez, 2013) and mathematical pedagogies that are equitable and culturally responsive (Aguirre, Mayfield Ingram, & Martin, 2013; Aguirre & Zavala, 2013; Gay, 2000; Greer et al, 2009; Moschkovich, 2013). Studies continue to show that equitable practices combined with high expectations, high-quality content, and strong family/community relationships have a positive effect on mathematics learning and achievement (Oakes, 2005; Kitchen et al., 2007).* (NCSM & TODOS, 2016, p. 3)

Essential Action

Ensure systems of continual collaborative, job-embedded professional learning to build teacher and leader capacity and increase efficacy.

In the first guiding principle, **Advocate,** mathematics leaders' actions support the development of culturally relevant mathematical tasks within curriculum leadership. To support teachers and teams in their use of rich mathematical tasks, leaders need to select professional learning opportunities and structures with a dual focus: (1) to increase knowledge of mathematics content and instructional practice and (2) to focus on culture, historical contexts, and language as well as other culturally relevant pedagogies in mathematics learning. Mathematics leaders structure ongoing allotments of time for teacher professional development learning and planning.

Changing practices requires focused attention on confronting and changing beliefs. A clearly articulated and comprehensive professional learning plan with clear and measurable outcomes is an important step in developing an equitable mathematics program and will be explored further in the Structures Supporting Teacher Professional Learning section of this chapter. Next up, we explore the strategic planning needed to support continuous professional learning focused on equitable instruction.

What Gets Planned Gets Done!

A mathematics leader's *commitment to others* includes a commitment to the action of creating a long-range strategic plan to improve all aspects of mathematics programming. For example, at the secondary level, teachers need opportunities to teach all levels of mathematics within a grade band. The practice of assigning more experienced teachers or those with status to teach upper-level courses such as calculus and assigning less experienced teachers with less status to freshman-level courses or courses designed for students who may need additional learning resources generates inequities for students and is no longer acceptable. Instead, effective leaders focus on the long-term goal of teacher professional growth and support in the teaching of mathematics content so that all students have access to effective teachers. In other words, effective leaders work to detrack teaching assignments as well as student placement.

Essential Action

Ensure sustainability through engaging all stakeholders in systemic, long-range strategic planning for all teaching and learning improvement initiatives.

Effective mathematics leaders continue to collaborate and seek support from a variety of stakeholders, including parents, experts in language acquisition, and special education, as new policies are developed and others evaluated. Leaders ensure research-informed practices are implemented with fidelity and ineffective practices are examined for the purpose of improvement or elimination. Finally, leaders continue to analyze prerequisites and support structures that lead to detracking pathways and success in community college and university nonremedial mathematics options. Promising practices for detracking course pathways are emerging from the San Francisco Unified School District (NCSM, in press; NCTM, 2019) and the Charles A. Dana Center at the University of Texas at Austin's work with higher education math pathways and statistics courses (Charles A. Dana Center, 2019).

For the sake of student and teacher learning, teaching assignments should balance grade-level and course-level teaching load. Additionally, to benefit both teacher growth and student learning, every few years, teaching assignments should change to include different levels and courses. When teacher assignments vary, teachers grow their toolbox of instructional strategies and deepen their understanding of the curriculum and how mathematical ideas progress and connect and students are the benefactors.

Further, having several teachers teach a particular grade or sections of a course fosters a sense of shared responsibility for student success. For example, the mathematics department at a charter school in Houston met at the end of one year to study student performance data and plan for the next school year. In an effort to improve their Algebra I student performance, each teacher agreed to teach one section of Algebra I so that they could collaborate in the course planning and work together to help all of their students succeed.

Recognizing the benefits to student and teacher learning, effective mathematics leaders create teaching assignments with the following attributes:

1. Teacher assignments vary across grades and course levels.
2. Over time, teachers are assigned different courses and grades.

Figure 2.6 contains actions that mathematics education leaders may facilitate when detracking students and teachers (NCSM, in press). Examine the actions and reflect on what you are currently doing in your leadership role. Record some thoughts about potential next steps that you could take to detrack students and teachers.

Detracking Action	What We Are Currently Doing	Additional Steps We Could Take
Ensure schools have aligned standards, instructional frameworks, and pedagogical guidance needed for heterogeneity in courses.		
Ensure that teachers and leaders are provided with professional development to develop the conditions needed for heterogeneity in their courses.		
Universities, colleges, and community colleges should continue to refine admissions language to better match the context and expectations of the CCSSM or other provincial or state curriculum standards.		
Broaden the purposes for teaching high school mathematics (including college and career readiness, professional success, understanding and critiquing the world, and experiencing the wonder and beauty of mathematics).		
When possible, balance teaching assignments at the high school level so that teachers have both upper-level and entry-level mathematics courses.		

Figure 2.6. Mathematics Leadership Actions for Detracking

Visit www.mathedleadership.org/EAresources to download a free reproducible version of this figure.

In order to maximize the impact of leadership efforts designed to improve mathematics teaching and learning, leaders must first focus on achieving access and equity in mathematics programming. A focus on teaching and learning, absent the discussion and professional learning focused on implicit bias, culturally responsive instruction, deficit mindsets, or social justice, is destined to fall short of truly attaining the shared goal that *each and every student can and must learn mathematics.* We will dig deeper into professional learning with a focus on equity later in this chapter.

Mathematics education leaders must intentionally design and implement structures that support access and equity, including providing opportunities for a *commitment to self* through self-reflection; a *commitment to colleagues* with an equity focus, featuring culturally relevant tasks, in professional learning; and a *commitment to others* through long-range, systemic planning, including detracking. Failing to do so sends a signal to some stakeholders, including teachers and students, that they are less important than others. In the next section, leaders will examine those structures critical to developing effective mathematics teaching and learning for each student.

Big Idea 2: Structures Supporting Teachers' Work With Students

Effective mathematics leaders ensure that targeted structures are implemented and monitored, resulting in every student having access to a rich and rigorous mathematics curriculum taught by collaborative teachers who are highly knowledgeable and skilled at mathematics instruction and assessment. Teaching and learning is the work of every leader in education. While it may be obvious to say that every decision and action of a mathematics education leader is about teaching and learning, without intentional structures and processes in place, it becomes extremely difficult, if not impossible, to ensure that the "right" work is being accomplished.

Myriad types of structures and processes are present within the complex operations of a district and school. Structures identified with teaching and learning understandably overlap with other supporting structures identified in equity and access, professional learning, and evaluation. The structures of particular focus in this section are those that are most directly aligned with what is happening around teachers and students that connect to curriculum, instruction, and assessment.

Curriculum and Student Intensification

Essential Action

Ensure mathematics learning for all students through organizational structures, time and resource allocation, and systemic supports that are aligned, intentional, and equitable.

In Chapter 1, we explored the difference between intended curriculum and implemented curriculum. In Chapter 2, we define the structures needed to close the gap between the intended, implemented, and eventually attained curriculum. Mathematics leaders, as they attend to the *commitment to self* in their sphere of influence, need to design and implement structures that ensure a mathematics curriculum that is coherent and meaningful in its design, reflective of state and national standards, and implemented by teachers.

As mentioned in Chapter 1, mathematics curriculum is not simply the textbook. "A curriculum is the program used to help students meet the standards, including instructional materials, activities, tasks, units, lessons, and assessments" (NCTM, 2014a, p. 70). Using NCTM's robust definition of curriculum, the importance of the work of the mathematics leader around curriculum is clearly visible. Our use of the term *curriculum program* is not intended to be one particular set of instructional materials; rather, it is the collection of instructional resources from which teachers select as they curate learning experiences for students. Curriculum development and implementation directly and profoundly impacts the mathematics learning of each and every student. The mathematics leader must ensure that structures are in place that result in a coherent, standards-aligned mathematics curriculum that is *implemented* by teachers and *learned* by students.

Instructional Resources

Curricular and instructional resources that are selected for use by teachers and students must also reflect research-affirmed best practice in the mathematics classroom, and at the same time, the allocation of these resources to students and teachers must be done in ways that are equitable.

"Instructional materials and resources are any concrete objects, digital devices, digital and print media, and support materials for lesson planning that enhance the teaching and learning process" (NCSM, 2014a, p. 33). Instructional materials are a requisite component of teaching and learning mathematics. Whether calculators, digital tablets and apps, manipulatives, or textbooks, instructional tools are used to engage students in making sense of the mathematics they are learning. Access to a variety of tools teaches and reinforces student learning, giving learners multiple ways to approach sense-making while encouraging flexibility of thinking in the selection and use of different tools. Teachers are the ones who engage students in the use of instructional materials and tools, so it is imperative that teacher teams be part of the process to identify and select instructional resources that frame the teaching and learning of mathematics.

Leaders who are the most impactful and intentional in their work do not assume that teachers know how to use instructional materials and resources in ways that have a positive effect on student learning of mathematics. Instead, effective mathematics leaders provide teachers and teacher teams with professional learning to best understand why and how best to use particular resources and tools for student learning.

Effective leaders of mathematics must:

1. Develop a process to identify and select high-quality instructional materials.
2. Engage teachers in the identification and selection of instructional materials.
3. Provide teachers with the professional learning needed to effectively use instructional materials.
4. Ensure allocation of instructional materials is done in ways that are equitable to benefit each and every student.

NCSM published a Curriculum Analysis tool that can also be used to support teachers and teams engaging in analysis of their instructional resources. Visit www.mathedleadership.org/EAresources to download the Materials Analysis Tool.

After the instructional resources are developed or selected, effective mathematics leaders focus on designing and implementing structures to support student learning through the use of these resources.

Figure 2.7 shows leadership actions that help design and implement structures supporting teachers' mathematics curriculum knowledge along with evidence of those actions. Use this tool to reflect on the current status of those leadership actions and identify possible next steps.

Actions Supporting Teachers' Mathematics Curriculum Knowledge	Evidence of Action	Current Status	Possible Next Steps
Ensure teachers and teams understand how and why focus, depth, and coherence make a mathematics curriculum effective.	Teachers and teams have a range of available opportunities to develop an understanding of the mathematics curriculum knowledge that relates to what their students are expected to learn.		
Develop and deepen understandings of learning progressions of key mathematical topics within a grade and across grades.	Teachers and teams routinely plan and implement lessons and units that reflect a coherent focus on important mathematical ideas.		
Organize the provincial or state curriculum content expectations for each grade or course into feasible teaching guides that link content standards, big ideas, and instructional resources.	Teacher and teams make consistent and effective use of regularly updated grade-level or course teaching guides.		
Create opportunities for teachers to investigate the curriculum at their grade level and across grade levels to fully understand the curriculum they expect students to learn.	Classroom observations consistently reveal accurate, appropriate, and effective sequencing of the mathematics being taught.		

Figure 2.7. Leadership Actions Supporting Teachers' Mathematics Curriculum Knowledge

Visit www.mathedleadership.org/EAresources to download a free reproducible version of this figure.

Collective Response to Student Learning

Curriculum is designed and implemented such that every student has access to a curriculum that is rigorous, relevant, and attainable. For some students, this may mean additional support or intensification. Intensification is defined as "a systemic effort to address the contextual needs of your students in learning on-level content" (Lynn & Briars, 2011, Slide 11). Many school systems refer to their systems of support or intensification as Response to Intervention (RtI). The bedrock of every RtI system is assurance that "instruction of the core curriculum using evidence-based practice [is] taught with fidelity so that all students are mastering the curriculum" (NCSM, 2013). When students are taught the curriculum in ways that support student sense-making and conceptual understanding of important mathematical knowledge and skills, most students will achieve the identified learning targets.

Discussed in more detail in NCSM's position paper on Tier 1 instruction (NCSM, 2013, p. 1), the mathematics leader puts systems and processes in place, resulting in mathematics classrooms that exemplify research-affirmed mathematics instructional and assessment practices such that most students are able to learn. However, for about 10% to 15% of students, additional support beyond effective classroom instruction is needed to attain the implemented curriculum and reach learning benchmarks.

Effective mathematics leaders create a "systematic method of identifying, defining, and resolving students' academic and social-behavior difficulties using collaborative, school-wide, problem solving approaches" (NCSM, 2015, p. 1). The mathematics leader convenes an RtI team, consisting of mathematics teachers and other school professionals with critical skills and knowledge of child development in general and of individual students specifically (e.g., counselors, psychologists, dean). Using clear and responsive criteria for identifying students who are not making adequate process, the mathematics leader and RtI team regularly monitor multiple and varied data points to collaboratively determine students in need of intervention and what actions are required to best intervene.

Mathematics leaders organize school-based problem-solving teams to use data systems to identify students who need additional time and support and create targeted interventions to meet those needs.

As mentioned in the first guiding principle, creating and leading a shared vision for mathematics is the starting point. The vision includes mathematics teaching and learning that supports learning for all students using a standards-aligned curriculum and evidence-based instructional practices. The first-best instruction includes differentiation to meet the needs of specific students, and teachers and teams monitor student learning through a combination of universal screeners, formative assessments, and progress monitoring in ways that are timely and responsive. Then, mathematics leaders organize school-based problem-solving teams to use data systems to identify students who need additional time and support and create targeted interventions to meet those needs. We will explore more ways to monitor student learning in the fourth guiding principle in Chapter 4.

Curriculum resources and student intensification plans are two ways that effective leaders can ensure mathematics learning for all students through organizational structures, time and resources allocation, and systemic supports. These actions, which support a *commitment to self,* are anchored in the foundational **belief** that all students can, and will, learn meaningful mathematics. There are additional actions that support teaching and student learning, which can be viewed through the imperative of *commitment to colleagues.*

Investment in Building Collective Efficacy

Teacher collaboration is an effective way to enhance teaching and learning in a school setting. Effective mathematics leaders harness the power of teacher collaboration as a part of their *commitment to colleagues* as a commitment to action supporting the building of teacher and leader capacity and increased efficacy. NCSM's position paper, *Improving Student Achievement by Learning Effective and Collaborative Teams of Mathematics Teachers* (NCSM, 2007), highlights key research showing how teacher collaboration has the potential to powerfully impact student learning. Allowing teachers to work in isolation results in vast inequities for student learning as a result of individual teacher decisions about curriculum, instruction, and assessment. When teachers collaborate, they bring together their collective experience and knowledge, including understandings about content, curriculum, and student learning. Together, teams identify what content and skills are essential, when and how to assess student learning, and how to respond based on analysis of assessment data. As teachers share data, knowledge, skills, and stories of students, the team develops a collective responsibility for the learning of mathematics by all students.

Essential Action

Ensure systems of continual collaborative, job-embedded professional learning to build teacher and leader capacity and increase efficacy.

Conditions Supporting Collaboration

Effective mathematics leaders are well aware that simply putting teachers together does not necessarily result in meaningful and impactful collaboration. The investment in resources required for teacher collaboration demand that leaders create and protect structures that require teacher collaboration, including dedicated time and clear expectations for how teams work with one another and the purpose and focus of their collaborative work. The end goal of collaborative teams is student learning of the curriculum for every student. Another equally important product of collaboration is the professional learning that happens when teachers focus on student learning. It is, therefore, imperative that the leader monitors the work of teams so as to identify professional learning needs of teachers, to provide feedback for ongoing growth, and to celebrate successes.

Figure 2.8 describes essential conditions and actions effective leaders of mathematics employ to support and encourage meaningful professional collaboration. Use Figure 2.8 as a reflection tool to identify strengths and opportunities for growth.

Conditions for Meaningful Collaboration	Next Steps
A culture of collaboration is the norm and that teacher isolation is not permitted.	
Teams are provided dedicated collaborative time and expectations for the work and outcomes of collaborative teams are clear, monitored, and celebrated.	
The primary work of teacher teams is to examine student learning data, reflect on instructional practice, and share best instructional practices.	
Teams have the collective responsibility to develop and implement a response to students who have not learned as well as to those students who have learned.	

Figure 2.8. Reflection on Conditions for Meaningful Collaboration

Visit www.mathedleadership.org/EAresources to download a free reproducible version of this figure.

Structures for Collaborative Work

Effective leaders of mathematics design and implement structures prioritizing collaborative professional learning in order to improve student achievement by allocating time, support, and resources for professional learning communities (NCTM, 2014a). This work represents a significant commitment of resources, including human resources, funding, and time (Kanold & Larson, 2012); the benefits of this collaborative study and problem solving emerge through improved student learning.

To this end, it is critical to articulate and model expectations for how collaborative teams work effectively together. Kanold and Larson (2012) identify five critical collaborative team areas that impact effectiveness and require leadership support:

- Participation—leaders must convene and support teacher teams with "a common curricular, instructional, or assessment focus about which to collaborate" (p. 14).
- Commitments—leaders must clearly communicate and model expectations for collaboration while facilitating the development of team norms that promote openness, vulnerability, respect, a willingness to share mistakes, and productive analysis.
- Leaders—formal and informal team leaders must facilitate effective collaboration while pressing the group for clarity, understanding, and collective implementation; structures and supports must equip these team leaders to effectively lead their teams.
- Agendas and meeting minutes—purposeful agendas acknowledge that time is valuable and facilitate a more efficient use of time; minutes capture the actions and decisions the team has made, as well as responsibilities for future work.
- Team time—leaders must ensure that necessary, sufficient, and consistent time is available and used for collaboration; this time must be within teachers' contractual workday.

Teachers and teacher teams, those who *implement* the curriculum, are essential to any and all structures and processes involving curriculum, including adoption, development, review, revision, and implementation of curriculum. In the process of interacting with the curriculum through collaborative discussions and processes, teachers and teacher teams negotiate collective meaning, share instructional approaches, develop assessment strategies, and explore the use of instructional materials, such as manipulatives and technology.

In other words, teachers and teams who interact with curriculum in focused and structured ways are developing their knowledge and skills of content, pedagogical content knowledge, and knowledge of student learning of mathematics. When teachers are denied the chance to work with curriculum, other than in the classroom, a leader has lost vital professional learning opportunities for teachers.

Use Figure 2.9 to identify possible structured opportunities for teachers and teams to engage in meaningful interactions with curriculum. What other structures do you use that could be added to Figure 2.9?

Structure	Artifacts and Evidence of Engagement
Built in weekly time for a grade-level or content-based team that includes regular attention on curriculum alignment, implementation, and attainment.	• Common assessments and scoring agreements • Clearly defined proficiency and rubrics • Lesson discussions about task implementation and/or cultural relevance
Collaborative teams to unpack curriculum standards to create a shared understanding of their intent.	• Clearly defined learning targets to share with students • Better understanding of the mathematical rigor of each standard
Collaborative teams to create a proficiency map (curriculum map) that illustrates the alignment and focus of standards connected to unit plans.	• Developing a pacing guide for implementation of the curriculum. • Completed unit plans
Teachers and leaders use the curriculum guide and/ or proficiency map to focus team assessments.	• Regular review of attainment of curriculum through data analysis of common assessments.

Figure 2.9. Possible Structured Opportunities

An effective mathematics leader's *commitment to colleagues* includes creating and supporting structures that intentionally generate teacher collaboration. Structures that support teachers' collaborative work also extend to engaging stakeholders outside the classroom, school, or district.

Engaging Stakeholders

Leadership does not happen in a vacuum. Effective mathematics leaders recognize and embrace the powerful impact of multiple voices contributing to vision, values, goals, and collective commitments toward a mathematics program that works for every child. Diversity of voices is a crucial element whether among district or school educators or inclusive of school families. This is why school leaders create structures that give voice to different perspectives. Effective mathematics leaders honor a *commitment to others* by committing to actions that intentionally include all stakeholders, including families, business partners, and community partners, in strategic planning for teaching and learning improvement initiatives.

Essential Action

Ensure sustainability through engaging all stakeholders in systemic, long-range strategic planning for all teaching and learning improvement initiatives.

Effective leadership sustains continuous improvement through goal setting and constant monitoring of those goals. The act of goal setting happens at multiple levels in a district or school organization. No matter the level of goal setting, a leader recognizes the power in compiling a diverse and collaborative team to determine goals and the subsequent action steps. Diversity of a collaborative team capitalizes on the fact that each stakeholder brings with her or him unique perspectives, experiences, and a set of knowledge and skills. When it comes to teaching and student learning, it is essential that the collaborative team is given a level of autonomy to set student learning goals that are meaningful to the team (NCSM, 2007) using disaggregated data by student population. This will be explored more in Chapter 4.

Families often choose neighborhoods or specific schools and districts because of the school system. Parents and guardians rightfully place a high value on their children's education. Because of the trust and expectations that families have for the education of their children, mathematics leaders seek to create partnerships and various ways to engage families and other community members in dialogue about the school mathematics programs. Cognizant that many adults have antiquated, misinformed, or unproductive beliefs about the teaching and learning of mathematics, including beliefs about what content is important, who should have access to which courses, or what mathematics instruction looks like, leaders are purposeful and intentional about creating structures to engage families in dialogue to inform about research-affirmed practices.

Use the tool in Figure 2.10 to consider how you, in your leadership role, address the given ways that help ensure sustainability of the continuous improvement of teaching and learning.

Strategy	How do we use this strategy? How might we use this strategy?
Create district and school teams that use data to set SMART goals and subsequent action steps.	
Ensure that progress toward goals is continuously monitored and revised as data analysis calls for.	
Develop mechanisms to communicate and inform families and the community about mathematics curriculum and research-affirmed instructional practices.	
Invite families and the community to be partners in dialoguing about changes and implementation of the school mathematics programs.	

Figure 2.10. Strategies to Support Engaging Stakeholders

Visit www.mathedleadership.org/EAresources to download a free reproducible version of this figure.

Mathematics education leaders must design and implement structures that support teaching and learning as they relate to classroom interactions among teachers and students. These structures include a commitment to self via careful selection of instructional resources and student intensification plans, a commitment to colleagues through conditions and structures supporting collaborative work, and a commitment to others by focusing on continuous improvement of existing structures affecting all stakeholders. In the next section, we will dive into structures crucial to developing and supporting professional learning for teachers and leaders of mathematics.

Big Idea 3: Structures Supporting Teacher Professional Learning

> *NCSM believes that intensive and sustained high-quality professional development for all teachers and leaders of mathematics is essential to improve student achievement.* (NCSM, 2007)

Effective leaders of mathematics design and implement structures that center professional learning in order to improve student achievement. To this end, leaders of mathematics must ensure that every teacher engages in continuous and impactful mathematics professional learning in order to improve their practice (NCSM, 2008). In this section, we will present actions that mathematics education leaders can take to design professional learning structures, provide a focus on collaborative professional learning, and develop and maintain a long-term professional learning plan.

Designing Professional Learning Structures

Essential Action

Ensure mathematics learning for all students through organizational structures, time and resource allocation, and systemic supports that are aligned, intentional, and equitable.

Consider all of the mathematics professional learning you have experienced. What was your best experience? What was your worst experience? Those experiences are worth keeping in mind as you exercise your *commitment to self* and design professional learning structures for teachers and leaders in your sphere of influence. Teacher growth focused on improving student achievement will take many forms to leverage the diversity of skills and strengths that teachers bring to the work. Wiliam (2016) observes that "the aim of professional development is not to make every teacher into a clone of every other teacher, but rather to support each teacher in becoming the best teacher he or she can be" (p. 168). Professional learning must support teachers, taking them from where they are toward that best teacher she or he can be, so as to maximize students' learning of mathematics.

First, it is important to dismantle notions of *developing* and *training* teachers. "Teachers do not need to be *staff developed*—something that is done to them. Teachers do need to participate in ongoing professional learning—an act of continual participation and action with others in the context of their workplace" (NCSM, 2007, p. 2). Effective leaders of mathematics center professional learning within their work while also designing and implementing structures that center professional learning within the organization.

Prioritizing professional learning involves modeling, engaging in, and leading professional learning as an ongoing *process* and not an event (Kanold & Larson, 2012). Structures, resources, and supports either focus on professional learning or expose it as a marginal activity within the organization (Loucks-Horsley, Stiles, Mundry, Love, & Hewson, 2010). For example, schedules that prohibit professional learning during the school day act as a major obstacle to job-embedded, collaborative practice. A lack of clarity and coherence in an organization's professional learning plan also represents a marginalized view of professional learning. Leaders of mathematics focus on professional learning through an active, continual commitment to implementing long-term, consistent priorities with a coherent focus that minimizes distractions and "initiative fatigue."

Designing Effective Professional Learning

Loucks-Horsley et al. (2010) find that effective professional learning experiences for mathematics teachers are driven by well-defined understandings of effective classroom teaching and learning, are collaborative and focused on a learning community, use the same instructional strategies that leaders want teachers to use, and support teachers in leadership roles. NCSM (2007) identifies key elements of intensive and sustained high-quality, content-based professional learning as elements that

- reflect a well-defined vision of quality mathematics and effective classroom practices;
- provide opportunities for teachers to enhance their mathematical knowledge and skills;
- improve teachers' classroom practice and lesson planning;
- are situated in the context of the teachers' work environment;
- provide time and resources for teachers to work in collaborative teams;
- are sustained and focused on student learning;
- promote high-quality, content-based teaching and address equitable outcomes for all students; and
- allow time for mathematics teachers to practice and reflect on various methods for teaching and representing mathematics content.

Leaders of mathematics design high-impact professional learning that focuses on student learning, analyzes the decisions of teachers and leaders, and attends to people and practices, not programs (Reeves, 2010). Effective professional learning focuses on improving student learning by observing, analyzing, and advancing the teaching practices within an organization. Additionally, effective professional learning is intensive and sustained, relevant to the needs of teachers and students, and focused on applying, practicing, reflecting on, and reinforcing teaching practices that move student learning forward (Reeves, 2010).

Ensuring Equity Through Professional Learning

Professional learning must include opportunities for teachers to confront and critically examine their biases and beliefs about who can and cannot learn mathematics (Loucks-Horsley et al., 2010).

When designing and facilitating professional learning, leaders can consider the critical reflection questions in Figure 2.11. The answers to these questions, and their implications, help expose attitudes and beliefs that may be preventing some students from accessing high-quality mathematics teaching and learning.

Reflection Question	Response and Implications for Next Steps
Does the content of the professional development experience include opportunities for teachers to examine and challenge their beliefs about who can learn and how diverse groups of students learn best?	
Does the content of the professional development experiences include the issues of equitable opportunity for all students to learn science and mathematics and participate in careers in science and mathematics?	
Is access to the professional development experience equitable? Is this opportunity available to all, or does it favor people in certain locations, with certain lifestyles, and from certain cultural, gender, or racial groups?	
Does the design of the professional development invite full engagement and learning by participants?	

Figure 2.11. Reflection Questions to Guide Equity Through Professional Learning
Source: Adapted from Loucks-Horsley et al. (2010, p. 135).

Visit www.mathedleadership.org/EAresources to download a free reproducible version of this figure.

A structural focus on equity involves attending to the ways that professional learning challenges or perpetuates conditions, biases, and behaviors that marginalize certain students and teachers.

Structural Considerations

Leaders of mathematics ensure professional learning impacts teachers' practice and students' learning through the use of organizational structures, time and resource allocation, and systemic supports. Professional learning that facilitates proficiency in teaching mathematics requires substantial resources, including the critical resource of time. The National Research Council challenges us to consider the time investment required to engage in professional inquiry and growth.

> *If teachers are going to engage in inquiry, they need repeated opportunities to try out ideas and approaches with their students and continuing opportunities to discuss their experiences with specialists in mathematics, staff developers, and other teachers. These opportunities should not be limited to a period of a few weeks or months; instead, they should be part of the ongoing culture of professional practice.* (National Research Council, 2001, p. 399)

The need for repeated opportunities to explore, discuss, apply, and learn iteratively represents the *substantive professional interaction* (Leinwand, 2012) required to grow teaching practice individually and collectively.

Organizations prioritize professional learning structure and safeguard substantive time *within the school calendar* for formal and informal professional learning opportunities such as peer observations, team teaching, professional sharing, classroom videoing and analysis, lesson study, issue-focused discussions, grade-level meetings (within and between schools), action research, course meetings, common planning, peer coaching, case study, mentoring, and collaborative grading (Kanold & Larson, 2012; Leinwand, 2012). Professional collaboration must take these many forms so that faculties "become dynamic communities of learners rather than just assortments of teachers and students working in the same building or district" (Leinwand, 2012, p. 92).

Districts and schools find creative ways to restructure time within the school calendar and instructional day in order to center professional learning (Loucks-Horsley et al., 2010).

- Released time—releasing teachers from their regular instructional time with students
- Restructured or rescheduled time—formally altering the overall instructional time, day, year, or teachers' schedules
- Common time—creating common noninstructional time for teachers
- Better-used time—reducing meetings, using meetings for professional learning, reallocating time from traditional professional learning experiences to job-embedded experiences
- Purchased time—paying teachers for work beyond their regular hours

An organization's central emphasis on professional learning will be reflected in its substantive, creative, and consistent investment of time. A sustained focus on professional learning, accompanied by an investment of time, demonstrates to teachers and other stakeholders that continuous and ongoing learning is truly valued by the organization (Kanold & Larson, 2012).

Implementing Structures

Effective leaders of mathematics design and implement structures that center collaborative professional learning in order to improve student achievement. NCSM (2007) recommends a set of actions that leaders can take in their work toward this goal. Use Figure 2.12 to determine which actions you are currently taking and which actions you might include in your work.

Actions for Structures of Professional Learning	Current Actions	Next Steps
Engage all mathematics teachers in a shared process for developing a clear vision of effective classroom learning and teaching of mathematics.		
Create professional learning goals based upon a local analysis of student mathematics learning and achievement.		
Provide intensive mathematics teachers' professional learning in order to allow time for practice, reflection, feedback, revision, and institutionalization for continuous improvement of mathematics instruction.		
Design mathematics professional learning activities that model effective classroom practices and lesson planning.		
Create mathematics professional learning opportunities for teachers to learn various approaches to teaching mathematics content.		
Create professional learning opportunities that engage mathematics teachers as learners, strengthening learning community cultures.		
Ensure that mathematics professional learning activities engage mathematics teachers in problem solving, integrating new learning, and collegial collaboration.		
Ensure that professional learning requires mathematics teachers to focus on all students learning important mathematics.		
Inform school and district administrators regarding the necessary policies and practices that will adequately support mathematics professional learning, linking it within the context of the educational system.		
Require administrators to participate in mathematics professional learning on an ongoing basis.		

Figure 2.12. Reflection on Leadership Actions

Visit www.mathedleadership.org/EAresources download a free reproducible version of this figure.

Designing professional learning structures supports mathematics education leaders' efforts to ensure mathematics learning for all students through organizational structures. These structures should be aligned to the shared vision for the school or district, anchoring them in the foundational belief of **vision.** Additionally, a plethora of research suggests that effective professional learning is collaborative. In the next section, we will explore ways in which leaders may center collaborative professional learning.

Focusing on Collaborative Professional Learning

> *Mathematics education leaders are responsible for leading teachers out of private practice into a collaborative working culture focused on making thoughtful and consistent decisions about curriculum, instruction, and assessment that will meet the unique needs of all students while at the same time helping students develop deep and connected mathematical understanding.* (NCSM, 2008)

Essential Action

Ensure systems of continual collaborative, job-embedded professional learning to build teacher and leader capacity and increase efficacy.

Effective leaders of mathematics design and implement structures that center collaborative professional learning in order to improve student achievement. The grain size for deep change in collective practice is the teacher team (Kanold & Larson, 2015). As such, leaders honor their *commitment to colleagues* when they design and implement structures that support collaboration, implementation, accountability, and growth within teacher teams.

Effective professional learning must be continuous and collaborative. It must be a *way of doing and being* within a school, rather than a set of events to attend. Prioritizing collaborative professional learning is a culture shift for many schools when interdependency becomes the norm where isolation used to permeate. Collaboration cannot be optional, as the costs of isolation are too high and the loss in collective expertise is too acute (Kanold & Larson, 2012). Continuous, collaborative professional learning leverages the strengths of all teachers in order to improve student learning for all students.

Job-embedded learning opportunities provide for purposeful, vital collaboration between teacher teams (Barnes & Toncheff, 2016). When teacher teams collaborate effectively, they are able to analyze and address complex issues in teaching. Kanold and Larson (2012) describe a "new paradigm for professional development [that] envisions mathematics teachers and other specialists collaborating interdependently to deepen their knowledge of mathematics pedagogical content and competencies, and expects *action on that knowledge* with application to practice" (pp. 8–9). Effective leaders of mathematics support collaborative teacher teams' growth by focusing on deliberate practice, which includes focused implementation, expert coaching, feedback, accurate self-assessment, and the opportunity to immediately apply feedback (Reeves, 2010). Job-embedded, collaborative professional learning is uniquely positioned to support the action-oriented work of collective inquiry and experimentation.

Figure 2.13 contains actions that leaders can take to foster collaborative professional learning (NCSM, 2007). Use the figure to reflect on which actions are already ongoing and which ones you might be able to implement in your leadership role.

Action	Current Actions	Possible Next Steps
Establish high-performing collaborative teacher teams and monitor the work of those teams. They should empower teacher teams to take actions that embody the shared values of the mathematics program by replacing norms of isolation with norms of collaboration.		
Make the purpose of the collaboration explicit; work deliberatively to build trust among the mathematics teachers and provide training for the team communication skills necessary to function together successfully.		
Help each teacher team to identify student achievement gaps in the grade or course level they teach and address the inequities caused by mathematics teacher isolation, privatization, and independent decision making.		
Radically monitor and celebrate teacher team performance, paying attention to the results in improved student achievement. Encourage discussions of work and successful initiatives to be an ongoing public endeavor.		
Collaborate with administration and other faculty to support the work and needs of the mathematics teacher teams during the contracted school day.		
There are intentional opportunities for teacher leaders to grow as leaders.		
Take responsibility for elevating and highlighting mathematics student achievement goals.		
Provide the current research and best practice evidence of the power of teacher teams to impact student performance.		
Help teachers learn to acquire culturally responsive pedagogical strategies and infuse, when possible and appropriate, culturally situated contexts and connections specific to the culture and communities of the students they teach and implement them in their instruction.		
Allow teachers to take responsibility to establish their own student achievement goals.		
Expect all teachers of mathematics to collaborate in a professional manner with integrity and honesty, using caring and respect.		
Not tolerate any form of teacher isolation and require all teachers to honor the decisions of the team as it applies to research-based best practices in mathematics curriculum, instruction, and assessment.		
Mandate that collaborative teams of mathematics teachers design required intervention programs targeted for struggling students early in the progress of each semester of the school year.		

Figure 2.13. Actions Supporting Collaborative Professional Learning

Visit www.mathedleadership.org/EAresources download a free reproducible version of this figure.

Ensuring that professional learning occurs collaboratively is an important essential action that effective leaders take. Doing so tends to the *commitment to colleagues* as well as affords the opportunity to critically reexamine **belief** structures and build **relationships** in the spirit of the four foundational elements of the *Framework for Leadership in Mathematics Education.*

Developing a Coherent and Focused Professional Learning Plan

Effective leaders of mathematics honor their *commitment to others* when they commit to actions that engage stakeholders in developing a coherent and focused professional learning plan. Additionally, effective leaders of mathematics communicate the importance of effective mathematics professional learning and its link to improved student learning. Effective leaders of mathematics engage stakeholders in developing a coherent and focused professional learning plan.

Essential Action

Ensure sustainability through engaging all stakeholders in systemic, long-range strategic planning for all teaching and learning improvement initiatives.

Shared Commitment and Support for Professional Learning

In order for mathematics initiatives to succeed and result in improved student learning, they must have expansive support from all stakeholders. This support is often contingent upon clear communication that results in a thorough understanding of an initiative, including its necessity, intended outcomes, and implementation process.

Leaders can garner public support for mathematics professional learning initiatives by increasing the awareness of effective mathematics teaching and learning, as well as effective mathematics professional learning (Loucks-Horsley et al., 2010). When members of the public, including families, business communities, and postsecondary education communities, understand the *why* of professional learning for mathematics teachers and leaders, they are more likely to support the *what* of professional learning.

To that end, effective mathematics leaders leverage partnerships with all stakeholders to develop a coherent, multiyear professional learning plan designed to improve mathematics teaching and learning for all. This plan must include the professional learning focus, what professional learning initiatives will support and yield these outcomes, who will provide and participate in various professional learning experiences, how the plan connects to the school's vision for exemplary mathematics teaching and learning, a timeline for learning and implementation, and an overview of the goals for each year (Barnes & Toncheff, 2016).

Teachers, leaders, and students are best served by a professional learning plan and process that focuses on deep, consistent implementation of a few critical practices (Reeves, 2010). The opportunities to engage in learning, applying, and implementing these practices should be numerous, differentiated, and sustained, but the critical emphasis of these professional learning experiences should be focused and coherent. We know the importance of differentiating instruction both for student learning and professional learning, including for teacher experience and the unique needs of teacher teams, individual teachers, and campus to district goals, is no different. By recognizing the diversity of skills educators bring to a school, organizations can design professional learning experiences that leverage and build on these strengths (NCSM, 2014b).

The professional learning plan should identify resources (including human resources, funding, and time) for professional learning. Public support for professional learning must include support for:

- Time and funding for professional learning
- Support for substitute teachers in classrooms for release time for teachers
- Support for days when school is closed for professional learning
- Teachers attending conferences and professional learning events at distant sites (Loucks-Horsley et al., 2010)

To this end, you must lead the effort to reconceptualize how to best use school time to center professional learning. Leaders can also garner support by modeling and conveying the importance of continual improvement as a form of professional accountability (Wiliam, 2016). When educators internalize and communicate the growth intended from professional learning processes, stakeholders see the critical effects this learning has on teacher practice and student learning.

Selecting a Professional Learning Focus

An effective professional learning program has a coherent and clearly articulated focus that permeates the work of the organization. Selecting this focus is "a complex dynamic in which the school culture filters, or interprets, data through its norms, values, beliefs, and theories to define problems and attribute causation" (Joyce & Showers, 2002, p. 63). The selection and refinement of this focus must be undertaken by a representative group of stakeholders invested in the mathematical teaching and learning of all students. Additionally, this professional learning focus must align with the school's clearly articulated vision for exemplary mathematics teaching and learning (Barnes & Toncheff, 2016).

Once a clearly articulated professional learning focus is determined, organizations should implement and support professional learning opportunities that build toward these intended outcomes (Wiliam, 2016). By focusing on *what* areas of practice in which a school is looking to improve, leadership teams can best determine *how* this growth would best be accomplished. The content (or goal) of professional learning should determine the process (or structure) of that learning.

Mathematics leaders, in conjunction with teachers and other stakeholders in a school or across a district, form a leadership team that manages the needed professional learning. Site-based or district-based leadership teams must carefully and critically attend to issues of content, pedagogy, accountability, assessment, and beliefs when designing professional learning.

Use the essential questions in Figure 2.14 to help guide discussions among your leadership team as you collaboratively create a professional learning plan.

Essential Questions for Professional Learning Planning	Responses
What mathematical content knowledge for teachers and leaders is necessary to meet our student achievement goals?	
What specific instructional strategies do we want to ensure all teachers and leaders fully understand and employ (strengthening pedagogy)?	
What part of our vision for exemplary teaching and learning do we not meet (strengthening accountability)?	
Does the faculty have the capacity to notice, analyze, and respond to students' mathematical thinking (strengthening assessment)?	
Do teachers' mindsets help or hinder student K–12 progression toward advanced mathematics coursework (strengthening beliefs)?	

Figure 2.14. Essential Questions for Professional Learning Planning
Source: Adapted from Barnes and Toncheff (2016).

After assessing current reality and identifying next steps, we move to developing the structures and means to monitor the implementation of the new learning.

Anchoring Professional Learning in Student Learning

Designing and providing professional learning in and of itself is not enough. Effective leaders of mathematics analyze the impact of professional learning on student learning through shifts in instructional practice (Joyce & Showers, 2002). Therefore, it is important for leaders to design measures that tangibly measure professional learning outcomes. One way to do so is to design and implement structures that foster collective accountability and monitoring of implementation aligned with the professional learning focus. Teachers, teacher teams, and leaders collectively set a target for implementation by defining what success will look like, collect data on an ongoing basis to evaluate the effectiveness of their implementation, and use these data to identify and troubleshoot obstacles to success (Joyce & Showers, 2002). Leaders play a critical role in this work as they facilitate teams capturing and monitoring their own teacher learning outcomes (NCSM, 2014b).

To ensure autonomy and support for professional learning, it is critical that leaders communicate the research evidence that sustained, intensive, collaborative professional learning improves student achievement (Loucks-Horsley et al., 2010). We must clearly identify, define, and describe what professional learning looks like at the local site, "increasing public awareness of and support for teachers' professional development" (Loucks-Horsley et al., 2010, p. 127) so that stakeholders can identify and support these critical practices. This includes "conveying the importance of teachers' ongoing learning outside of the classroom and emphasizing how this enhances student learning. Changing perceptions about what professional development 'looks like' and how it benefits student learning can increase understanding of its importance among families and community members" (p. 127).

NCSM (2014a) provides several ways that leaders can observe connections between professional learning and student learning outcomes. Use Figure 2.15 to reflect on current actions to align professional learning and student actions to identify possible next steps.

Characteristics of Professional Learning Aligned to Student Learning	Current Actions	Possible Next Steps
Outcomes for professional learning are related to mathematical content knowledge, pedagogical knowledge, and specialized content knowledge.		
Peer dialogue and feedback support classroom endeavors.		
Collegial sharing of learning experiences occurs inside and outside the school.		
All teachers and team members participate in professional learning opportunities and are expected to apply new learning to practice.		
Teachers develop a coherent plan to document professional goals and progress toward them.		
Teams predicate learning opportunities on what they know from research-affirmed best practices for professional development in mathematical content knowledge, pedagogical knowledge, and specialized content knowledge.		
Teachers have data-informed conversations with colleagues about what is working and what is not working in order to ensure that all students are achieving.		
Educators celebrate and share individual and collective successes.		
Students are realizing measurable benefits (improved learning, improved social-emotional well-being, etc.) from teacher engagement in professional learning.		

Figure 2.15. Professional Learning Aligned to Student Learning

Visit www.mathedleadership.org/EAresources to download a free reproducible version of this figure.

The nexus between professional learning and student learning shows the intersection of a number of the foundational beliefs of this leadership framework. Effective professional learning focuses on the **belief** that all students can and will experience and learning meaningful mathematics. The pursuit of a shared **vision** is how teachers make change that positively impacts their students' mathematical learning. Professional learning, as with student learning, is **equitable** and addresses the learning needs of all educators. Because effective professional learning is collaborative, **relationships** are an essential part of how teachers learn and thus, in turn, provide more effective learning experiences for their students.

Summary of Essential Actions

Strategic **design** is at the heart of efforts to improve student mathematics learning. Systems and structures supporting access and equity, mathematics teaching and learning, and professional learning are essential to achieving a vision that supports every student receiving and every teacher delivering high-quality mathematics instruction every day (Barnes & Toncheff, 2016). Mathematics education leaders may design and implement structures supporting access and equity by

- providing opportunities for teachers and others to self-examine their existing beliefs and attitudes,
- leading with equity by ensuring that professional learning is equity focused and includes culturally relevant tasks so that teachers can see culturally relevant learning in action, and
- developing and deploying system, long-range plans, including those that seek to detrack existing student learning pathways.

Mathematics education leaders may design and implement structures supporting teachers and their work with students by

- leading teachers through curriculum and student intensification planning,
- building collective efficacy through conditions and structures that support collaborative work, and
- engaging stakeholders in ongoing reflection and continuous improvement of existing structures.

Mathematics education leaders may design and implement structures supporting professional learning by

- designing effective professional learning structures, including incorporation of equity and consideration of time, funding, focus, and a monitoring plan for implementation of professional learning;
- structuring professional learning so that it is collaborative in nature; and
- developing and refining a long-term professional learning plan that includes shared buy-in and connections to student learning.

Taking Action

Figure 2.16 is a reflection on the three imperatives aligned to the second guiding principle: **Design** and implement structures that support high-quality mathematics teaching and learning for every teacher. Complete the reflection to identify strengths and areas of possible growth.

Action	Reflection Questions	Comments
Imperative 1: *Ensure mathematics learning for all students through organizational structures, time and resource allocation, and systemic supports that are aligned, intentional, and equitable.*		
Provide opportunities for self-reflection on how equitable learning opportunities are currently being provided and how they could be made more available.	How do we currently provide equitable learning opportunities for ALL students? In what other ways could we provide more equitable learning opportunities for particular groups of students?	
Create and implement equitable structures for student intensification plans.	What systematic method is used to identify, define, and resolve students' mathematical and social-behavior difficulties? What structures are currently used to support student intensification?	
Design and implement structures that support professional learning that supports mathematics learning for all students.	What are the professional learning needs of our mathematics team?	
Design, implement, and refine a system of teacher evaluation that focuses on student learning.	To what degree does our current evaluation system focus on student learning? How might we improve observation practices so that the evidence collected and the formative feedback provided to teachers are aligned to research-affirmed best practice for mathematics teaching?	
Imperative 2: *Ensure systems of continual collaborative, job-embedded professional learning to build teacher and leader capacity and increase efficacy.*		
Embed culturally relevant practices into professional learning opportunities and incorporate contexts that are relevant to students in professional learning.	How might we use culturally relevant teaching practices to deliver teacher professional learning? What contexts could we use to support mathematical content in professional learning?	
Create and implement opportunities for teachers to collaborate.	How might we develop a master schedule that provides substantive time for job-embedded professional learning and collaborative planning?	
Ensure that professional learning is collaborative in nature.	What processes will be established to ensure that the designed team professional learning experience leverages the strengths of the full team?	
Anchor professional learning in a shared vision of high-quality mathematics teaching and learning.	What resources might be drawn upon to ensure that the team professional learning experiences reflect research-affirmed best practice?	
Imperative 3: *Ensure sustainability through engaging all stakeholders in systemic, long-range strategic planning for all teaching and learning improvement initiatives.*		
Collaborate with all stakeholders, including parent groups, community organizations, and experts in fields such as special education or English language acquisition.	How well do all of our students have access to meaningful grade-level content? What additional resources do we need so that all stakeholders can learn more about equitable learning opportunities for all students?	
Engage all stakeholders in order to ensure sustainability of the continuous improvement of teaching and student learning.	What experiences might we design so that stakeholders gain a clear picture of the current and desired state of mathematics teaching and learning?	
Develop a coherent and focused professional learning plan.	Who are the stakeholders responsible for the long-term design and facilitation of a professional learning plan? What tool will be used to gather data about the strengths and needs of participants of professional learning?	
Collaborate with all stakeholders on a cycle of continuous improvement for teaching and learning, curriculum development, and staffing.	How might we leverage the school improvement planning process (or districtwide strategic planning process) to set goals and monitor the progress toward goal attainment?	

Figure 2.16. Self-Reflection on Guiding Principle 2

Visit www.mathedleadership.org/EAresources to download a free reproducible version of this figure.

The essential leadership actions for the second guiding principle provide a roadmap for mathematics leaders engaged in improvement efforts. As you reflect on the list of actions, record areas of strength and weakness with the school, district, state, or province that looks to you for leadership. Develop a list of next steps, prioritizing each as near term or long term.

In this chapter, you learned more about the second guiding principle: **Design** and implement structures that support high-quality mathematics teaching and learning for every teacher. Effective mathematics leaders ensure both student and adult learning by structuring supportive conditions for job-embedded continuous learning. In the next chapter, we will examine the third guiding principle: how to **Empower** and nurture a culture of productive professionalism.

CHAPTER 3

Guiding Principle 3: Empower

Big Idea 1: Beliefs and Mindsets

Big Idea 2: Building Relationships

Big Idea 3: Culturally Sustaining Practices

Summary of Essential Actions

Taking Action

"Leaders become great, not because of their power, but because of their ability to empower others."

—John C. Maxwell

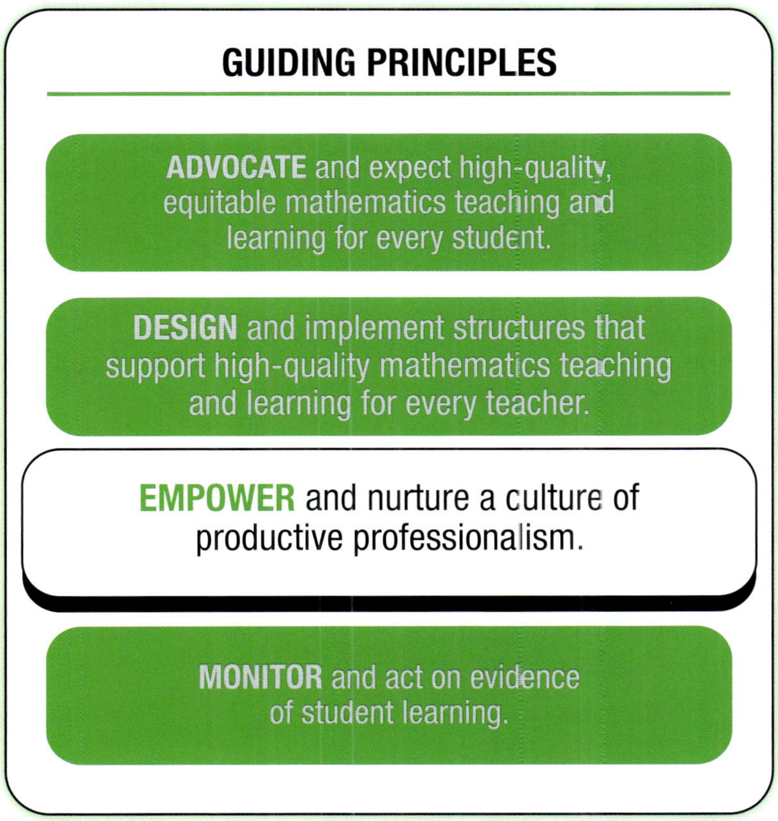

Guiding Principle 3: EMPOWER and nurture a culture of productive professionalism.

As we have learned, mathematics leaders **advocate** for and expect high-quality and equitable mathematics teaching. We also **design** and implement structures to support that high-quality teaching and learning. To accomplish those goals, mathematics leaders must **empower** teachers and nurture a culture of productive professionalism.

Mathematics education leaders strive to enable and foster a set of shared attitudes, values, goals, and practices that benefit all students. This guiding principle is founded on our foundational elements: beliefs, vision, equity, and relationships. Empowering teachers and nurturing a culture of productive professionalism requires educators and team members to examine their beliefs, create a common vision, attend to equity among students and teachers, and establish and nurture relationships among all stakeholders.

Consider the scenarios in Figure 3.1.

Scenario 1: LaKeasia, a young teacher early in her career, had the opportunity to participate in professional learning opportunities that resulted in her reexamining her personal teaching practices. With the support of the instructional facilitator at her school, she began to change the culture in her classroom. When visiting her classroom, one would see students working collaboratively on mathematical problems and tasks. Students shared and compared their strategies and processes, discussing which methods were most easily understood and which were most efficient. LaKeasia worked to identify and facilitate problems and tasks that provided opportunities to ensure all of her students experience and learn meaningful mathematics. The end of each lesson was a reflection on the lesson by the students themselves; they identified what they were doing well and new learning goals. Visiting this classroom was an amazing experience, and word spread to other districts in the area. They began requesting permission to have their teachers visit this classroom.
Scenario 2: A team of three Algebra I teachers decided they would volunteer to meet at their lunchtime once a week to examine student work products and other data to attempt to reflect and refine their practice to hopefully improve the achievement of their students. They found this so useful, they began meeting at lunch every day and often would stay late after school to continue their discussion and refinement of their plans for the next day. They consistently had student work products to share and began to identify instructional practices that were effective. They reported that traditional lecture demonstration instruction was often ineffective, and they began to search for ways to increase levels of student engagement. They were pleased to see the grades in their classes improve, and when the state-mandated assessment was given near the end of the year, scores were significantly better than the previous year. The geometry teachers in the same school were impressed by the results. The following year, there was a geometry professional learning team in the school. These teams were not mandated by the administration, but they were supported by the building administration and an instructional facilitator for the school.

Figure 3.1. Leadership Scenarios That Reflect Actions From the Empower Principle

These two scenarios show how teachers, individually or collectively, look for ways to nurture a culture of productive professionalism among students and teachers. Mathematics education leaders cultivate and empower a culture that supports those collaborative ideas. LaKeasia's story shows how she took charge of her own learning to restructure her own practice. The Algebra I team is a model of how teachers can take initiative together to study data and use the results to inform their own practice. In both scenarios, effective mathematics leaders created conditions and structures that empowered teachers to work together and change their practice to improve student learning.

In this chapter, we will examine the guiding principle, *empower and nurture a culture of productive professionalism,* and explore each of these in light of the three essential actions shown in Figure 3.2.

	Imperative 1 *Commitment to Self*	**Imperative 2** *Commitment to Colleagues*	**Imperative 3** *Commitment to Others*
EMPOWER and nurture a culture of productive professionalism.	**Essential Action** *Ensure assumptions, beliefs, expectations, and habits are examined in order to shape the school or department culture around teaching and learning mathematics.*	**Essential Action** *Ensure a culture of reflection, refinement, and action focused on continuous improvement in mathematics learning.*	**Essential Action** *Ensure students, teachers, and community partnerships are built upon meaningful engagement.*

Figure 3.2. Empower Guiding Principle From the Framework for Leadership in Mathematics Education

Within the empower guiding principle, the commitments to action among the different imperatives emerge as follows:

- Actions related to the first imperative, *commitment to self,* focus on how a leader works with educators in her or his sphere of influence to examine beliefs and mindsets, relationships, and cultural selves.
- Actions related to the second imperative, *commitment to colleagues,* focus on how a leader works with educators in adjacent spheres of influence, such as other teacher teams, other departments, on a campus, or within the district, to support beliefs and mindsets with a different type of professional knowledge, build relationships with colleagues, and value perspectives from multiple stakeholders.
- Actions related to the third imperative, *commitment to others,* focus on how a leader works outside of his or her immediate learning community in order to consider and inform the mindsets of other stakeholders, strengthen relationships with other stakeholders such as families and community organizations, and invite community partners to participate in the work of the school.

The **empower** imperatives show the importance of shared beliefs (foundational element), expectations, and habits among all stakeholders so that schools, districts, or organizations may create a culture of reflection, refinement, and action in order to continuously improve student learning outcomes. Effective schools ensure that teachers, students, and community partnerships are meaningful. This chapter is designed to help bring these imperatives to action by focusing on the following:

- Big Idea 1: Foundational assumptions and beliefs (mindsets)
- Big Idea 2: Building relationships
- Big Idea 3: Cultivating culturally sustaining practices for self, others, and the community

As a mathematics leader, what is your role in empowering and nurturing a culture of productive professionalism? Use the guiding questions in Figure 3.3 to reflect, based on your leadership role, on how you as a bold leader can support the three essential actions within the **Empower** guiding principle.

Guiding Principle 3: EMPOWER and nurture a culture of productive professionalism.	
Essential Action	**Guiding Questions**
Ensure assumptions, beliefs, expectations, and habits are examined in order to shape the school or department culture around teaching and learning mathematics.	In my leadership role: • Do I believe that all students can learn? • Do I understand actions that support developing a collaborative professional culture? • What are my beliefs, assumptions, habits, and expectations regarding teaching and learning?
Ensure a culture of reflection, refinement, and action focused on continuous improvement in mathematics learning.	In my leadership role: • How do I support the development of professional knowledge (e.g., mathematics content knowledge, pedagogical content knowledge, mathematics curriculum knowledge)? • Do our professional learning and other organizational structures encourage collaboration? • Do all students have access to challenging mathematical courses?
Ensure students, teachers, and community partnerships are built upon meaningful engagement.	In my leadership role: • How do we engage families and community members in mathematics teaching and learning? • In what ways do we include families, businesses, and community members in our work? • How do we include community interests in mathematics teaching and learning?

Figure 3.3. Leadership Roles With the Empower Guiding Principle

Visit http://mathedleadership.org/EAresources for a role-specific reflection tool with questions for each mathematics leadership role (teacher leader, coach, site-based leader, district leader, and state/provincial leader).

Now that you have reflected on your initial beliefs and your current situation, let's examine the key ideas that support the three essential actions aligned with the **Empower** guiding principle.

Big Idea 1: Beliefs and Mindsets

> *NCSM is a mathematics education leadership organization that equips and empowers a diverse education community to engage in leadership that supports, sustains, and inspires high quality mathematics teaching and learning every day for each and every learner.* (NCSM Mission Statement, 2017)

We need to nurture and empower a culture of productive professionalism in order to achieve the NCSM mission and vision. We must be ready to challenge our own thinking, including our own beliefs, expectations, and habits, and those of others in order to ensure "high quality mathematics teaching and learning every day for each and every learner." We need to develop, encourage, and embrace a growth mindset for ourselves, our colleagues, and the communities we serve. A growth mindset is a place that is open for revision and new thinking (Boaler, 2015; Dweck, 2016). It allows us to take risks and to develop our own bold leadership.

In this section, we will examine how attending to our mindset can contribute to empowering and nurturing a culture of productive professionalism. At the beginning of this chapter, you met three Algebra 1 teachers who decided to analyze student work samples in order to improve their teaching (Figure 3.2). These teachers developed a working theory of action that if they could collaborate, analyze data, and reflect on its implications, that they could improve their collective teaching and thus improve student achievement. A set of common beliefs and assumptions creates a mindset through which leaders, and the teachers who they support, view their work. A productive mindset is an essential part of empowering a culture of productive professionalism.

The Importance of Beliefs and Mindset

Essential Action

Ensure assumptions, beliefs, expectations, and habits are examined in order to shape the school or department culture around teaching and learning of mathematics.

We have read the fairy tales and seen the movies where the evil queen or the beautiful princess looks in a mirror to see what really exists. Similarly, a mindset that leads to the empowerment of a culture of productive professionalism starts with looking in a "mirror" so that we can see what initially exists regarding our own beliefs and attitudes. Effective mathematics leaders engage in a commitment to self when they work with those in their immediate sphere of influence to examine their own assumptions, beliefs, expectations, and habits in order to determine how likely they are to nurture productive professionalism. There are two key factors that mathematics leaders will employ as we empower a productive culture: bold leadership that helps teachers to challenge their existing assumptions and beliefs, as well as an examination of personal mathematical identities.

Bold Leadership Challenges Its Own Assumptions and Beliefs

NCSM President Connie Schrock (2017–2019) said, "I believe there is much we must do to support the teachers that we lead. Bold leadership goes beyond a typical job description" (NCSM, 2019d, p. 1). Lambert (2002) asserts that when teachers' beliefs change, they view their instructional practice in new ways. She goes on to suggest that the work of leaders is to engage teachers in sustained conversations about teaching and learning mathematics.

Mathematics leaders are responsible for creating opportunities and conversations that challenge teacher beliefs. We bring out the mirror! These conversations are part of the ongoing, regular dialogue between the leader and teacher, both informally and formally. Ignoring the power of beliefs and mindset or missing opportunities to shift beliefs and mindsets undermines the likelihood of change and improvement (NCSM, 2014b, p. 14).

Consider this quote from Dr. Brené Brown (2018):

> *I want to live in a world with braver, bolder leaders... we desperately need more leaders who are committed to courageous, wholehearted leadership and who are self-aware enough to lead from their hearts, rather than unevolved leaders who lead from hurt and fear.* (p. 4)

> **"I believe there is much we must do to support the teachers that we lead. Bold leadership goes beyond a typical job description."**
>
> **— Connie Schrock, NCSM President (2017–2019)**

How are you committed to lead with courage when it comes to engaging in critical conversations? A study by the Green Peak Partners and Cornell School of Industrial and Labour Relations (Flaum, 2010) found self-awareness to be the strongest predictor of overall success of leaders. The researchers found that leaders who were self-aware were stronger leaders. The leaders inspired trust, were able to set high expectations, and used strong relational skills and respect to support meeting the expectations. Self-awareness begins with asking yourself the hard questions. Consider the following questions:

- Do I have low expectations for certain populations?
- Do my teachers integrate social-emotional learning competencies into their mathematics instruction?
- Do all students have access to challenging mathematical courses?
- Are we reproducing patterns of privilege and disadvantage in our classrooms by being a gatekeeper of mathematics? (Noguera, 2018)

Have a courageous conversation with yourself first so you can have them with your teachers and leaders that you serve.

> **"Daring leaders who live into their values are never silent about hard things...Living into our values means that we do more than profess our values, we practice them."**
>
> **— Dr. Brené Brown**

Overcoming the inequities in mathematics education is an ongoing challenge and critical if we are going to live our values. "Daring leaders who live into their values are never silent about hard things...Living into our values means that we do more than profess our values, we practice them. We walk our talk—we are clear about what we believe and hold important, and we take care that our intentions, words, thoughts, and behaviors align with those beliefs" (Brown, 2018, pp. 184, 186). Instantiate Dr. Brown's advice about living into your values by being bold. Ensure examination of assumptions, beliefs, expectations, and habits in order to shape the school's or department's culture around teaching and learning of mathematics.

To boldly examine your current assumptions, beliefs, expectations, and habits and access and equity in mathematics, use the tool in Figure 3.4.

To what degree do you agree or disagree with these statements in the context of your school?

Agree				Disagree
1	2	3	4	5

1. Mathematics ability is a function of opportunity, experience, and effort—not of innate intelligence. Mathematics teaching and learning cultivate mathematics abilities. All students are capable of participating and achieving in mathematics, and all deserve support to achieve at the highest levels.
2. Equity is attained when students receive the differentiated supports (for example, time, instruction, curricular materials, programs) necessary to ensure that all students are mathematically successful.
3. Equity—ensuring that all students have access to high-quality curriculum, instruction, and the supports that they need to be successful—applies to all settings.
4. Students who are not fluent in English can learn the language of mathematics at grade level or beyond at the same time that they are learning English when appropriate instructional strategies are used.
5. Effective mathematics instruction leverages students' culture, conditions, and language to support and enhance mathematics learning.
6. Effective teaching practices (for example, engaging students with challenging tasks, discourse, and open-ended problem solving) have the potential to open up greater opportunities for higher-order thinking and for raising the mathematics achievement of all students, including poor and low-income students.
7. The practice of isolating low-achieving students in low-level or slower-paced mathematics groups should be eliminated.
8. All students are capable of making sense of and persevering in solving challenging mathematics problems and should be expected to do so. Many more students, regardless of gender, ethnicity, and socioeconomic status, need to be given the support, confidence, and opportunities to reach much higher levels of mathematical success and interest.

Figure 3.4. Productive Beliefs About Access and Equity in Mathematics Tool
Source: NCTM (2014a, pp. 63–64).

Visit www.mathedleadership.org/EAresources to download a free reproducible version of this figure.

Audre Lorde once said, "When I dare to be powerful, to use my strength in the service of my vision, then it becomes less and less important whether I am afraid." Bold leadership requires strength in the service of your vision. Your vision becomes more clear as you examine your personal mathematics identity.

Personal Mathematics Identity

As mathematics leaders build collective capacity, we are constructing our identities as mathematicians and leaders. As we develop meaning of mathematics leadership, we are also developing a knowledge of how to serve others. Our personal identities allow for bold leadership that embraces learning with courage, taking risks, and having a sense of possibilities. Our personal identity is forming through reflective interactions with others (Lambert, 1995).

Mathematics identity is defined as the dispositions and deeply held beliefs that students develop about their ability to participate and perform effectively in mathematical contexts (Aguirre, Martin, & Mayfield-Ingram, 2013). Teachers, peers, and parents can all exert an influence on the mathematics identities that students develop. A key consideration about mathematics identities is that they are strongly connected with the other identities that students construct and view as important in their lives, including their racial, gender, language, cultural, ethnic, family, faith, and academic identities (Aguirre et al., 2013).

Sami Briceño (2018) advises to be mindful of mindset messages you and your students communicate to yourselves and each other. Words have power. Are the words used as a vehicle to empower students or belittle them? All educators must reflectively ask themselves, "How have I influenced the mathematical identities of my students?" Reflection is the key to honing an educator's craft.

Use questions like the ones shown in Figure 3.5 to guide a reflection on mathematical identities.

Reflective Question	Response
What range of mathematics identities are expressed and performed by your students?	
What actions do you take to positively affirm your students' mathematical identities?	
What are some of the various identities that your students express and perform through the stories that they narrate in your mathematics classroom?	
How do you model positive mathematical agency and provide opportunities for students to demonstrate this?	
What additional family roles do your students take on that might contribute to their positive development in mathematics?	

Figure 3.5. Reflective Questions for Mathematical Identities

Visit www.mathedleadership.org/EAresources to download a free reproducible version of this figure.

Bold leaders build their capacity to sharpen the mathematical lens through an active engagement in ongoing professional learning.

Bold leaders build their capacity to sharpen the mathematical lens through an active engagement in ongoing professional learning. Bold leaders attend conferences, webinars, and online classes; participate in district/regional/provincial professional development; visit other schools; observe colleagues; collaborate with a mathematics mentor principal; join the mathematics community on Twitter; subscribe to blogs; read journals; and join national professional organizations such as NCSM: Leadership in Mathematics Education or National Council of Teachers of Mathematics (NCTM) and local affiliates of these two mathematics professional organizations. Both NCSM and NCTM maintain lists of current affiliates on their respective websites.

Effective mathematics education leaders self-examine their own beliefs and mindsets, including their own mathematical identities. They also create structures for those whom they lead, students, or other educators to do the same. Examining your own beliefs and mindsets honors your *commitment to self.* Designing and creating structures for your colleagues to reflect on their beliefs and mindsets, as well as take action to sharpen them, honors your *commitment to colleagues.*

Supporting Beliefs and Mindsets With Different Types of Professional Knowledge

> *Mathematics is a very broad and multidimensional subject that requires reasoning, creativity, connection making, and interpretation of methods; it is a set of ideas that help illuminate the world; and it is constantly changing. Math problems should encourage and acknowledge the different ways in which people see mathematics and the different ways in which people see mathematics and the different pathways they take to solve problems. When these changes happen, students engage with math more deeply and well.* (Boaler, 2015, p. xii)

Essential Action

Ensure a culture of reflection, refinement, and action focused on continuous improvement in mathematical learning.

To build a culture of reflection, refinement, and action, we must be open to new ideas and new ways of thinking. Mathematics education leaders honor their *commitment to colleagues* beyond their immediate spheres of influence when they create opportunities for professional learning and opportunities for teachers and leaders to collaborate to nurture alignment with the shared vision. Leaders also create opportunities to learn from colleagues through observations and discussions of the complex work of teaching in pursuit of that shared vision of high-quality mathematics teaching and learning that we explored in Chapter 1. Effective mathematics education leaders build a culture where each of their colleagues feels comfortable in confronting teaching problems head on and are willing to seek feedback from one another.

One way to work toward improving learning for all students is to connect learning with our students' senses of mathematical identity, as we described in the previous section. In the turn toward a commitment to a culture of refinement, reflection, and action, mathematics leaders should consider the different types of knowledge that mathematics teachers must possess. These types of knowledge are helpful in framing the substance being refined, reflected, and acted upon.

NCTM Past President Francis (Skip) Fennell said, "Effective principals also understand the importance of teachers' mathematical and pedagogical knowledge. They find ways to help teachers develop, maintain, and deepen this knowledge—perhaps through an analysis of state or local curriculum frameworks, targeted professional development opportunities, or formal coursework" (Fennell, 2007, p. 1). Bold mathematics leaders know, model, and support the pedagogical and content knowledge of their staff by leading by example.

There are several types of professional knowledge that effective mathematics teachers possess. Using Figure 3.6, reflect on the current professional learning and identify possible next steps needed to cultivate these types of knowledge.

Type of Knowledge	Current Opportunities: What Else Might We Need?
Mathematics content knowledge refers to a person's understanding of the skills, concepts, applications, reasoning methods, and connections within mathematics that are prerequisite to what is being taught, are aligned with what is being taught, and build from what is being taught (NCSM, 2014b).	
Pedagogical content knowledge refers to the critical knowledge that links specific mathematics content with effective mathematics instruction (NCSM, 2014b; Shulman, 1986). This knowledge is what enables a skilled teacher to determine the most effective instructional strategies for particular content.	
Mathematics curriculum knowledge is an understanding of how to sequence and organize the content for teaching (NCSM, 2014b). Typically, this sequence follows established learning progressions that are aligned with state standards and show how mathematics ideas flow from one grade or course to the next.	
Mathematical knowledge for teaching is "a kind of professional knowledge of mathematics different from that demanded by other mathematically intensive occupations" (Ball, Hill, & Bass, 2005, p. 16). This type of knowledge requires teachers to know deeply the interrelationships among mathematical concepts, skills, and procedures, much more so than someone who uses mathematics in their occupation, such as an engineer or accountant.	

Figure 3.6. Types of Knowledge

Visit www.mathedleadership.org/EAresources to download a free reproducible version of this figure.

Content knowledge is the "what," whereas pedagogical knowledge is the "how." Leaders must be aware of these pivotal concepts in order to address common misconceptions so students can progress. Establish an understanding that content and pedagogy are inseparable components of effective teaching. "Leaders must be held accountable for ensuring that teachers have the content knowledge and instructional skills to help students increase their mathematical understanding. Leaders must be able to model appropriate instruction, find resources, ask deep questions to ensure understanding, and guide teachers to improve their mathematics knowledge and repertoire of instructional strategies" (NCSM, 2014b, p. 23).

A critical examination of educators' beliefs and mindsets, as framed by the types of mathematical and curriculum knowledge they possess, helps mathematics education leaders empower and nurture a cycle of reflection, refinement, and action. Next up, we will examine how working with mindsets alongside families and the community supports a culture of productive professionalism.

Beliefs and Mindsets of Stakeholders

Let your children see you fail and recover. Let them fail in order to see it as a growth opportunity.

Effective mathematics leaders honor their *commitment to others* by extending discussions of beliefs and mindsets to all stakeholders. According to Dr. Stephanie O'Leary (2017), a clinical psychologist specializing in neuropsychology, failure is good for children on several levels. First, children who experience failure learn how to cope, a skill that is certainly needed in the real world. It also provides children the life experience needed to relate to peers in a genuine way.

Essential Action

Ensure students, teachers, families, and community partnerships are built upon meaningful engagement.

When students experience failure, it instills the need for hard work and sustained efforts, and it also demonstrates that hard work and effort are valuable even without the blue ribbon, gold star, or top score. Over time, children who have experienced defeat will build resilience and are more willing to attempt difficult tasks and activities because they are not afraid to fail. On the flip side, rescuing children too early sends a message that parents don't trust them. When parents and other adults are willing to allow students to struggle, the action communicates that adults believe students are capable of handling any outcome, good or bad (DesMarais, 2017).

Consider the following scenarios in Figure 3.7.

Scenario 1: A fifth-grade team is implementing a new curriculum with more conceptual representations of multiplication and division of fractions. The teachers are implementing a unit of instruction where they are trying to connect the relationship between multiplication and division of fractions. They want their students to understand and explain why the procedures for multiplying and dividing fractions make sense. They have received several questions from families, and they created some videos to send home with children. They also hosted a math night and the students led stations where they could play math games and model their new learning.
Scenario 2: A district mathematics leadership team has supervised the implementation of the new standards over the past 5 years. Student learning is stagnant, and they wanted to answer the question, "What might students have to say about efforts to improve mathematics classroom activities?" They decided the best way to answer the question was to interview students. They developed open-ended questions to understand students' perspectives on classroom instruction and environment. They interviewed a variety of students at each school, listened carefully, and recorded all of the comments. The data were then transcribed and shared with the school leadership teams to analyze and inform next steps. The perspective data informed the mathematics teachers and leaders about the quality of teaching and learning and issues of equity and access.

Figure 3.7. Building Partnerships Scenarios

These two scenarios are examples of how mathematics leaders can engage students and families. Dina O'Brien, a mathematics specialist from New Jersey, understands that all parents want to help their children succeed, so she asks them the following questions:

- How are beliefs about math being reinforced?
- What kinds of math learning goals are being encouraged?
- Are mistakes valued in the math learning process?

She provides parents with a few truisms to shift their everyday language signals that reflect a growth mindset, shown in Figure 3.8.

Consider mistakes as a first attempt at learning, not failures.
Encourage hard work.
Compliment new and different solving strategies (especially if families haven't been exposed to these solving strategies themselves!).
Refrain from criticizing alternative solving strategies that families might be unfamiliar with.

Figure 3.8. Shifts in Everyday Language
Source: Adapted from O'Brien (2019).

A better understanding of beliefs and mindsets helps mathematics education leaders empower teachers and nurture a culture of productive professionalism. This understanding begins with a self-examination of bold leadership and personal mathematics identities. Leaders honor the *commitment to colleagues* with an examination of the different types of professional knowledge that teachers hold. Leaders honor the *commitment to others* by including stakeholders outside of the schoolhouse. Focusing on beliefs and mindsets is often an internal action, whether it is internal to the self or internal to a campus or organization. We shift the focus to an external one by examining the building of relationships in the next section.

Big Idea 2: Building Relationships

One of the four foundational elements on which this book is based is *relationships.* Mathematics leaders need to work collectively and build working relationships among all stakeholders to implement and achieve our vision for mathematics teaching and learning.

In this section, we will examine how building relationships contributes to empowering and nurturing a culture of productive professionalism within the context of the three imperatives.

Examining and Developing Your Own Relationships

Essential Action

Ensure assumptions, beliefs, expectations, and habits are examined in order to shape the school or department culture around teaching and learning of mathematics.

Carr, Herman, and Harris (2005) describe mentoring, coaching, and collaboration as ways to build professional relationships. When effective mathematics leaders build professional relationships, they are honoring their *commitment to self* by leading within their own spheres of influence. Each type of relationship has its own purpose and skills.

- **Mentoring** relationships nurture retention, induction, and professional support and require skills such as observing and giving feedback, receiving feedback, and planning for teaching and learning.
- **Coaching** or peer coaching nurtures instructional improvement and professionalism and requires skills such as teaching or observing demonstration lessons and using description versus judgment when giving feedback.
- **Collaboration** takes the form of study groups, committees, teacher teams, or advisory councils and leadership groups. Collaboration nurtures professional growth, organizational change, and action planning and requires skills such as the use of protocols to guide collaborative inquiry.

Teaching is a social endeavor that requires relationships in order to be successful. Students learn best when they have a positive relationship with their teacher (Sparks, 2019). By extension, teachers learn and grow best when they have a positive relationship with their colleagues. Lieberman and Miller (2004) identify three transformational shifts that support teachers in their professional growth: moving from individualism to professional community, moving from teaching at the center to learning at the center, and moving from technical and managed work to inquiry and leadership.

Effective mathematics education leaders create structures that empower these transformational shifts, reducing teacher isolation by nurturing professional learning communities, shifting the instructional focus from teacher behaviors to student learning outcomes, and transforming the work of teaching from managerial tasks to shared inquiry. Using these structures (from Chapter 2), mathematics leaders support creating a professional culture while simultaneously developing relationships. Establishing strong relationships is key to these shifts.

Use Figure 3.9 to identify structures that currently exist and could be strengthened or new structures that you could create to help teachers and leaders cultivate strong professional relationships with their colleagues.

Goals for Structures	**How could we embed this in an existing structure?** **What new structures could we create to support this goal?**
Reduce teacher isolation by nurturing professional learning communities.	
Shift the instructional focus from teacher behaviors to student learning outcomes.	
Transform the work of teaching from managerial tasks to shared inquiry.	

Figure 3.9. Structures Supporting Building Relationships

Visit www.mathedleadership.org/EAresources to download a free reproducible version of this figure.

By starting with a focus on a shared vision of high-quality mathematics teaching and learning, teacher teams can identify specific shared values that pertain to their particular work as it supports the shared vision. Shared vision and values create a foundation from which teachers can then build their own relationship networks.

At the beginning of the chapter, you read a scenario about LaKeasia, a young teacher who attended a professional learning event with an open mind. She compared her assumptions, beliefs, expectations, and habits to the ones that were being advocated in the professional learning event. From this event, she was able to cultivate relationships that aligned with the shared vision and values within her school. Particularly, she cultivated relationships with her building level mathematics coach (instructional facilitator) and the state mathematics specialist who had organized the event. Because she had cultivated relationships with willing mathematics education leaders, she was able to leverage those relationships when she needed to.

Because the leaders with whom she forged relationships were willing to engage with her for her professional growth, she was able to develop a teacher-as-learner approach to incorporating new instructional strategies in her own practice. These leaders, through their bold leadership and support, created a teacher who will constantly learn new things and work to improve her practice.

Building relationships often begins with your own growth as a teacher or as a leader. As you mature as a teacher or leader, becoming more confident in your own work, you begin to look outward to develop relationships that benefit not only you but your colleagues as well.

Essential Action

Ensure a culture of reflection, refinement, and action focused on continuous improvement in mathematical learning.

Building Relationships With Colleagues

> *It can be daunting to be told to change our practice, to take risks, and to try something new. But when a colleague we know, whether next door or in another school far away, tries something different, it somehow becomes more concrete and, perhaps, achievable.* (Seeley, 2009, pp. 210–211)

NCTM Past President Cathy Seeley compares the development of great teaching to the development of great mathematical thinkers (Seeley, 2009). As mathematical thinkers build on the work of others, "teachers and leaders in school mathematics also build on the work of others every day, both those who are famous and those who are less well-known" (p. 210). Effective mathematics education leaders know this and, in honoring their *commitment to colleagues,* work hard to create structures, such as face-to-face or online professional learning communities that facilitate teachers in building on their work.

Knight (2011) describes a partnership approach to building relationships in the context of professional learning. He identifies seven principles that support professional learning as a partnership among stakeholders.

- Equality, in which leaders see themselves as a colleague, leading from the midst rather than the front
- Choice, in which teachers have authentic choices about which workshops to attend
- Voice, in which leaders recognize that professional learning values perspectives of all participants
- Reflection, where all participants in the learning process, including leaders, have the opportunity to make sense of what is proposed that they learn
- Dialogue, in which ideas are exchanged in productive ways
- Praxis, in which teachers have the opportunity to enact what they are learning
- Reciprocity, in which leaders and teachers alike learn from each other in an ongoing cycle of professional learning

Kanold (2017) frames relationships as alliances, which are mutually beneficial relationships requiring support and action from both directions. Alliances are powerful structures that allow teachers and leaders to collaborate, strengthen the authenticity of their professional learning communities, attack and reduce inequities, streamline professional decision making, strengthen a sense of relational intelligence, provide powerful opportunities for peer observation and feedback, and allow us to celebrate our successes.

Effective mathematics education leaders use these ideas to create structures that ensure collective reflection, refinement, and action that nurture a culture of productive professionalism. Leaders not only use structures to empower teachers in their care to support one another. Additionally, leaders turn to one another—their own colleagues—through professional organizations, local collaboratives, or other formal or informal groups. For example, in the Dallas–Fort Worth area, regional mathematics leaders meet monthly to ask one another questions, share ideas that they are using in their own districts, and tap the regional expertise in pursuit of shared solutions. Many of these leaders also attend meetings of the Texas Association of Supervisors of Mathematics twice annually in order to learn together and receive important information with state leaders. A subset of the regional leaders annually attends the NCSM Annual Meeting to learn alongside colleagues at the national level, bringing new ideas back to their home group. These professional networks are invaluable means of professional support for leaders.

Additionally, teachers and leaders are increasingly turning toward social media to build their networks. These digital relationships help to reduce geographic isolation, particularly for rural teachers and leaders who may not be able to travel hundreds of miles to meet regularly with colleagues. Structures such as #MTBoS (Math Twitter Blog-o-Sphere) and certain

Facebook groups, such as NCSM's Coaching Corner, provide places where teachers and leaders may ask one another questions, share successful (and unsuccessful) strategies, and learn in a virtual environment alongside one another. Gover (2017) points out that social media help teachers create professional learning networks (PLNs) that serve as an outlet for enhancing pedagogy, as teachers have access to colleagues across the country and around the world, as well as help foster collaboration among teachers. Of course, social media are not without risk, and teachers must become proficient at evaluating ideas and advice given to them in their professional learning networks. Effective leaders provide their teachers with strategies to build these virtual relationships and evaluate the quality of instructional resources, mathematical tasks, and pedagogical advice they are given in unfiltered forums that may not be adequately peer-reviewed.

Effective leaders of mathematics must act to create and sustain structures that will support a culture of meaningful and ongoing professional learning for all mathematics teachers. Moreover, this professional learning should be provided within the context of the teacher's local work environment, much like what we have seen with LaKeasia's story.

The professional development experiences are designed to model effective classroom practices and lesson design. The experiences are aligned to the expectations from Chapter 2 and feature various instructional approaches to teaching mathematics content. Teachers and leaders are engaged as learners and, in order to strengthen learning community cultures, exposed to activities focused on problem solving, integrating new learning, and collegial collaboration.

Effective mathematics education leaders create and refine structures, like teacher teams who use collaborative inquiry to strengthen their collective practice, that empower and nurture a culture of productive professionalism. In effective schools and districts, that culture extends beyond the school and includes other stakeholders, such as families and community partnerships.

Building Relationships With Others

Meaningful engagement is a cornerstone of partnerships among schools, families, and communities. Effective mathematics education leaders honor their *commitment to others* as they attend to relationships within the school or district as well relationships with external stakeholders.

Students' families and community members have a particular set of beliefs about what mathematics is and how it should be taught. There are certain expectations for the types of homework that children bring home and the types of skills that potential employees bring to their employers. As Kreisberg and Beyrenavand (2019) observe, tensions emerge when those expectations conflict with what is actually happening. When states revised their curriculum standards, including the adoption of Common Core State Standards for Mathematics, parent and family groups and social media exploded with misrepresentations of research-affirmed instructional practices, mostly revolving around elementary mathematics. Much of this tension could have been mitigated with effective communication and building stronger relationships where all community members' voices are included and honored.

Essential Action

Ensure students, teachers, families, and community partnerships are built upon meaningful engagement.

Maintaining communication with families is key for any school or teacher. DuFour and Eaker (1998) identify key elements of communication that support successful relationships between schools and families.

- Communication between the school and home must be regular, two-way, and meaningful.
- Communication is most powerful when it is timely.
- Communication protocols should not place excessive demands on teachers.
- Communication must be two-way.

Teachers and leaders can use a variety of strategies to include students' families in the learning that occurs in classrooms. Teachers like LaKeasia in the introduction of this chapter could collaborate in teacher teams to produce a weekly mathematics newsletter, either print or electronically, to send home with students, informing families about what their students are studying and how they are progressing in their learning. Sharing information and video clips about the instructional strategies, representations, and models that students are using to learn mathematics helps parents and family members better understand the evidence of student learning they are seeing in work products (Kreisberg & Beyrenavand, 2019). In Figure 3.7, you read about a fifth-grade team that created video clips about multiplication and division of fractions in response to questions they received from families. These video clips, along with LaKeasia's newsletter, are good examples of ways to include families in conversations about changes in instruction we experience in response to changing provincial or state curriculum standards.

Sheldon (2011) identified four characteristics of successful school–community partners, shown in Figure 3.10. Use these characteristics to reflect on your existing school–community partnerships. How well do they exhibit these characteristics? What additional steps could you take in your leadership role to include these characteristics in strengthening existing or to create new partnerships?

Characteristic	Description	Reflection
Employ teamwork	Form an Action Team for Partnership with teachers, administrators, parents, community members, and, if possible, students	
Write annual, goal-oriented action plans	Create an annual action plan that aligns with the campus or district improvement plan	
Use a multidimensional definition of involvement or engagement	Parenting and supportive family environments, two-way communication about school programs and student progress, volunteering at school, learning at home, decision making, and collaborations with community	
Evaluate partnership practices	Include ongoing end-of-year evaluations of partnership programs and practices with reflections and action steps for improvement	

Figure 3.10. Reflection on Characteristics of Successful School–Community Partners

Effective mathematics education leaders are able to build relationships that honor the *commitment to self* in your current leadership role, the *commitment to colleagues* in formal and informal organizations, and the *commitment to others* by including parents and the community in campus or district relationships. Next, we will look at a third way leaders may empower and nurture a culture of productive professionalism, the use of culturally sustaining practices.

Big Idea 3: Culturally Sustaining Practices

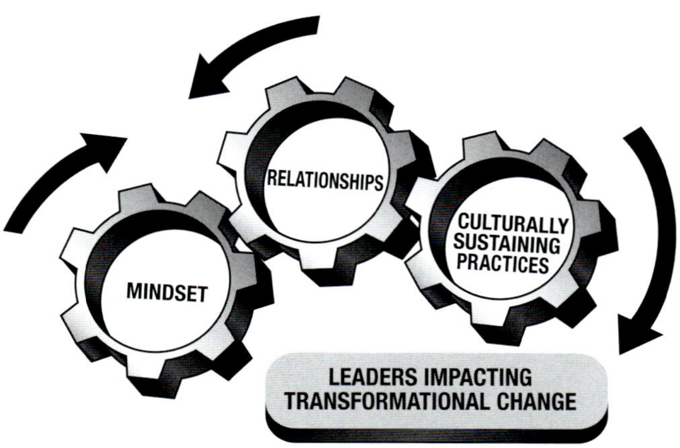

Figure 3.11. Tenets of Culturally Sustaining Practices

In previous sections, you have read about the importance of building relationships and establishing a growth mindset. Figure 3.11 demonstrates how the tenets of what we are calling *culturally sustaining practices* work in tandem with building relationships and establishing a growth mindset to help leaders empower and nurture a culture of productive professionalism.

Cultural sustaining practices apply to an examination of self, to valuing the perspectives of colleagues, and to engaging the community. Culturally sustaining practices are grounded by culturally sustaining pedagogy, which seeks "to perpetuate and foster—to sustain—linguistic, literate, and cultural pluralism as part of schooling for positive social transformation" (Paris & Alim, 2017, p. 1). Culturally sustaining practices seek to promote attributes for leading social change in schooling for positive transformation, while pushing back on biases, stereotypes, deficit thinking, and preconceived notions. As well, culturally sustaining practices seek to utilize the pluralistic nature of our society to guide transformations. These transformations include *empowering and nurturing a culture of productive professionalism,* as well as school practices and policies, school culture, professional development, and student growth (academically, socially, and emotionally), extending as far as your imagination can take you.

When thinking of cultural pluralism, most think of race, gender, religion, language, and so on. However, cultural pluralism is much broader. Other factors, such as community assets, generational cultures, and political viewpoints, bring dynamic insights into the conversation. Consider generational culture, particularly regarding Generation Z (Gen Z, people born between 1997 and 2003), as an example. Currently, Gen Z-ers are permeating schools as students, teachers, and leaders. This generation is considered self-aware, self-learning, self-reliant, and entrepreneurial. Other characteristics include persistent, realistic, innovative, and carrying a desire to make things happen—like betterment of the environment (Merriman, 2015). Gen Z-ers have not known a world without the Internet. They can obtain information with the swipe of their finger.

Therefore, educators and leaders must be more than disseminators of facts and algorithms; they must be facilitators of critical thinking and problem-solving tasks that incorporate hands-on learning experiences. Because of the richness of variance individuals bring to the discussion, building relationships, having a growth mindset, and incorporating culturally sustaining practices are necessary to empower and nurture a culture of productive professionalism. In the following sections, we explore what culturally sustaining practices mean in relation to self, colleagues, and the community.

Educators and leaders must be more than disseminators of facts and algorithms; they must be facilitators of critical thinking and problem-solving tasks that incorporate hand-in learning experiences.

Revealing Your Cultural Self

With respect to culturally sustaining practices, mathematics education leaders honor their commitment to self when they commit to actions encouraging teachers and leaders in their spheres of influence to engage in self-reflection. "Who is your cultural self?" is a question you must ask. Look beyond your physical self and into your beliefs, bias, assumptions, habits, and expectations while also remembering that part of your cultural self is that of a mathematician/mathematics educator and leader. "Emerging leaders are construing and interpreting themselves as they construct meaning and knowledge with others. Their sense of personal identity allows for courage and risk, low ego needs, and a sense of possibilities. Personal identity has been forming in reflective interactions with others. They seek not only to explain, but to listen and to understand" (Lambert, 1995, p. 49). It is when you conduct a deep-dive into your cultural self that space is opened for you to challenge deficit thinking and to invite others into candid conversations.

Essential Action

Ensure assumptions, beliefs, expectations, and habits are examined in order to shape the school or department culture around teaching and learning of mathematics

Aristotle mused that knowing yourself is the beginning of all wisdom. In that spirit, self-awareness begins with asking yourself the hard questions. Figure 3.12 contains some guiding questions to structure this reflection.

Questions for Reflection
• What experiences (challenges and supportive) have shaped me? • Have these experiences developed into asset thinking, a growth mindset, or other ways of thinking? • Do I have low expectations for certain populations? • Am I, as Dr. Pedro Noguera (2018) says, reproducing patterns of privilege and disadvantage in our classrooms by being a gatekeeper of mathematics?

Figure 3.12. Guiding Questions for Cultural Self-Awareness

An African proverb tells us, "Unless you know the road you've come from, you cannot know where you are going." The first imperative for the guiding principle states that bold mathematics leaders *ensure assumptions, beliefs, expectations, and habits are examined in order to shape the school or department culture around teaching and learning of mathematics.* Thus, a true look in the mirror is necessary.

Valuing Perspectives

When someone with the authority of a teacher describes the world and you are not in it, there is a moment of psychic disequilibrium, as if you looked into a mirror and saw nothing. (Rich, 1986, p. 178)

Essential Action

Ensure a culture of reflection, refinement, and action focused on continuous improvement in mathematical learning.

In this statement, Adrienne Rich referred to learners. However, her statement also applies to teachers and leaders when that someone in authority is a leader. Lambert (2002) suggests that the work of leaders is to engage teachers in sustained conversations about teaching and learning mathematics. This means leaders have to actively listen and allow varying perspectives to emerge by creating opportunities and conversations that challenge teacher beliefs.

A few guiding questions are as follows:

- How do we integrate social-emotional learning competencies into their mathematics instruction, or do all students have access to challenging mathematical courses?
- How do we address concerns of educational inequities immediately?
- What measures do we have in place to provide opportunities for antibiased and pluralistic expressions of the curriculum?

These are bold questions that require bold leaders to "open themselves to and other teachers to new ideas, risk imaginatively, and enthusiastically inspire those they lead with a desire to learn and grow together" (NCSM, 2008, p. 56). It is easy to fall into the trap of deficit thinking. However, leaders who utilize culturally sustaining practices reject ideas that focus on what is "broken," and they do not lose sight of the many assets available within faculty and students.

To capitalize on the vast insights and experiences of your colleagues, time and space must be provided for these crucial conversations to occur in a safe space. Bold leaders must also ensure equity (one of the foundational elements) is present during these conversations. If these things are not present, colleagues may not fully participant, leading to valued perspectives being lost. The *TRU Math Conversation Guide* (Baldinger, Louie, & the Algebra Teaching Study and Mathematics Assessment Project, 2016) says, "Our most meaningful learning occurs when we interact with others… They offer perspectives on our work that differ from our own. And they respect our intelligence, skill, and intentions—as well as our need to continually grow." These culturally sustaining practices ensure a culture of reflection, refinement, and action focused on continuous improvement in mathematical learning and assist in empowering and nurturing a culture of productive professionalism.

Bringing in Partners

Essential Action

Ensure students, teachers, families, and community partnerships are built upon meaningful engagement.

An African proverb that we hear frequently is "it takes a village to raise a child." As mentioned with all of the third imperatives, we cannot forget to engage families and community members as partners. The National Staff Development Council (cited in NCSM, 2008, p. 3) says, "Quality teaching in all classrooms necessitates skillful leadership at the community, district, school and classroom levels." In a culture of productive professionalism, we cannot impact transformative change in mathematics education alone. To gain the most significant results, effective leaders not only include teachers but also engage students, families, and the community in the process.

Mathematics leaders may have the preparation of a professional educator, but students, families, and community bring experiences that can have a tremendous impact on student learning in mathematics education (NCSM, 2008). With the same culturally sustaining practices of inclusiveness of colleagues and critical examination of self, the community must be afforded the same opportunities to engage.

Bold leaders seek opportunities to collaborate with students, families, and the community in creating meaningful partnerships. One may think involving the community is difficult, but if you keep your eyes open to unlikely partners and open-minded, what emerges may surprise you.

Consider the interview, a portion of which follows, with Dr. Dante Edwards, Assistant Principal, who is at an urban secondary school in Georgia that serves 100% African American students. He has partnered with parents and a local institution of higher education to create opportunities for students to learn computer science. As you read the interview, consider what actions were needed to develop this unique opportunity for students.

> *The goal of the higher education program was to promote computer science to underrepresented youth. The program had a rocky beginning and the institution sent an email to local secondary schools inviting them to participate. I grabbed the opportunity quickly. We partnered to develop ways to market the program to parents, teachers, and students. We wanted to put students first. Therefore, we removed all barriers that would prevent students from enrolling. We even changed our master schedule. We only had a few teachers who were certified to teach computer science, therefore we engaged our mathematics faculty. Mathematics teachers co-taught with professors from the institute during the first year and gained more autonomy and confidence in the years to follow such that they are now independently teaching the course. Students, parents, and teachers in this area pride themselves on "being the first… being unique." Therefore, they immediately bought into the idea of learning anime and app development.*

It was through the collaborative examination of the culture of the school community of wanting to put students first that the mathematics leader chose to take advantage of the opportunity. Since the generational culture has a propensity toward technology, the partnership was created to benefit the students, school, and then the district. The culturally sustaining practices in bringing in partners were the flexibility of the school, gathering multiple perspectives, and making sure there was time provided for reflection and revisions.

Use the tool in Figure 3.13 to consider a few guiding questions to ask in engaging the community in meaningful partnership that build on empowering and nurturing a culture of productive professionalism.

What knowledge and experiences does the community offer?
What risks and opportunities do I take to engage the community?
How do I support community engagement?

Figure 3.13. Guiding Questions for Community Engagement

Empowering and nurturing a culture of productive professionalism through culturally sustaining practices can have an enormous amount of positive implications. These implications are due to the elimination of bias and deficit thinking and also due to valuing pluralistic experiences of self, others, and the community. Possible implications include

- increased teacher morale,
- increased strength of instruction,
- improvements in mathematical engagement and learning,
- potentially decreased teacher burnout and turnover,
- additional opportunities in STEM careers for students,
- expansions of ideas beyond one school,
- community building, and
- accountability.

The key to creating a culture of productive professionalism that embodies culturally sustaining practices is *inclusive collaboration*. Figure 3.14 is a visual representation of this statement. Stakeholders have a responsibility to openly share, but it is the bold leader who facilitates.

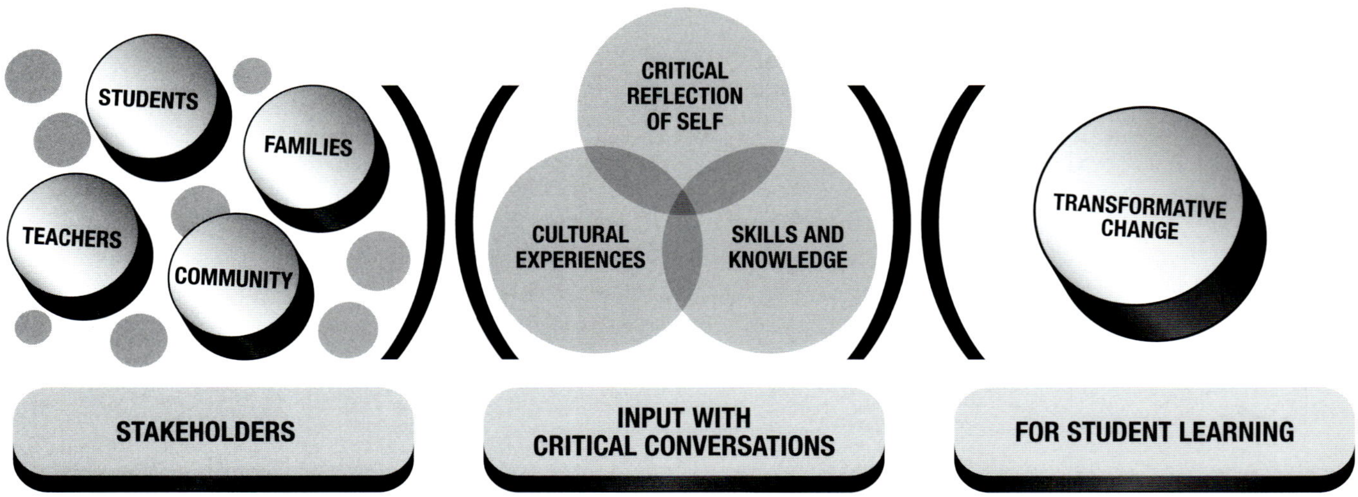

Figure 3.14. Inclusive Collaboration

The bold mathematics leader skillfully creates a safe environment where perspectives can be candidly expressed and where all stakeholders are valued in creating transformative change that will impact the teaching and learning of mathematics.

Summary of Essential Actions

Empowering teachers and nurturing a culture of productive professionalism is at the heart of efforts to improve student mathematics learning through improved teaching practices. Kanold and Larson (2015) argue that the best strategy to improve student learning is through professional learning community structures, and the ideal grain size is the teacher team. These two critical ideas support leaders' work in empowering teachers and nurturing a culture of productive professionalism. Mathematics education leaders may nurture productive professionalism through beliefs and mindsets by

- providing opportunities for teachers and others to self-examine their existing beliefs and attitudes,
- examining the types of existing knowledge and need for additional knowledge within the teacher or leadership team, and
- fostering and strengthening relationships among all stakeholders, including families and communities.

Mathematics education leaders may nurture productive professionalism through building relationships by

- leading teachers through an examination of their own relationships;
- creating structures that support collaborative, collegial relationships at all levels; and
- forging meaningful partnerships with families and community groups and organizations.

Mathematics education leaders may nurture productive professionalism through culturally sustaining practices by

- fostering critical self-reflection of cultural awareness,
- creating an environment in which all values are respected, and
- honoring the proverb, "It takes a whole village to raise a child" by including and inviting families and community partners to participate in the work of the school.

Taking Action

Figure 3.15 is a reflection on the three imperatives aligned to the third guiding principle: **Empower** and nurture a culture of productive professionalism. Complete the reflection to identify strengths and areas of possible growth.

Action	Reflection Questions	Comments
Imperative 1: *Ensure assumptions, beliefs, expectations, and habits are examined in order to shape the school or department culture around teaching and learning of mathematics.*		
Reflect on existing mindsets, including educators' personal mathematical identities.	What do I believe about how students best learn mathematics? How do I support students in the development of their mathematical identities?	
Encourage teachers to build relationships that support their own professional learning and mindsets.	What opportunities do I provide for teachers and leaders that I serve to meet and network with others whose relationships might be beneficial?	
Foster an environment that supports teachers as they examine and identify key elements of their own cultural selves.	What opportunities do I provide for teachers and leaders that I serve to examine their own cultural selves?	
Imperative 2: *Ensure a culture of reflection, refinement, and action focused on continuous improvement in mathematical learning.*		
Create a culture of collective responsibility for student learning.	How do I support teachers in cultivating a belief in a shared responsibility for student learning?	
Cultivate a professional culture that rewards collaborative professional growth and learning.	What structures do we have that encourage teachers and leaders to collaborate? What structures do we need to consider to increase collaboration?	
Encourage educators to value the perspectives of others in pursuit of shared culturally sustaining practices.	What opportunities do we have to compare our perspectives to those of others, seeking commonalities and appreciating differences?	
Imperative 3: *Ensure students, teachers, families, and community partnerships are built upon meaningful engagement.*		
Acknowledge and cultivate the role of productive struggle as a part of the learning process for all stakeholders, including students, teachers, parents, and community.	How do leaders in my campus or organization model productively struggle for teachers and thus students?	
Create a culture that encourages partnerships with all stakeholders, including teachers, families, and community.	What partnerships do we currently have with families and community organizations? What others would better support our students?	
Situate the school culture in the context of the community culture.	How does the school culture reflect the community culture? Are all aspects of the community culture represented in our school culture? If not, why not?	

Figure 3.15. Self-Reflection on Guiding Principle 3

Visit www.mathedleadership.org/EAresources to download a free reproducible version of this figure.

In this chapter, you learned more about the third guiding principle: **Empower** teachers and nurture a culture of productive professionalism. Relationships, a foundational NCSM belief, are essential to this part of the work of mathematics education leaders. Central to these relationships is the idea of collaboration among colleagues, particularly in teacher teams (NCSM, 2007). As well, effective mathematics leaders focus on examination and careful cultivation of beliefs and mindsets, intentional ways to build collaborative relationships, and purposefully implementing culturally sustaining practices. In the next chapter, you will dig deeper into the final guiding principle, monitoring and acting on evidence of student learning.

CHAPTER 4
Guiding Principle 4: Monitor

Big Idea 1: Monitoring Student Learning at the Classroom Level

Big Idea 2: Monitoring Student Learning in Teacher Teams

Big Idea 3: Monitoring Student Learning at the District Level

Summary of Essential Actions

Taking Action

"Assessment is a process that should help students become better judges of their own work, assist them in recognizing high-quality work when they produce it, and support them in using evidence to advance their own learning."

—NCTM, Principles to Actions, p. 92

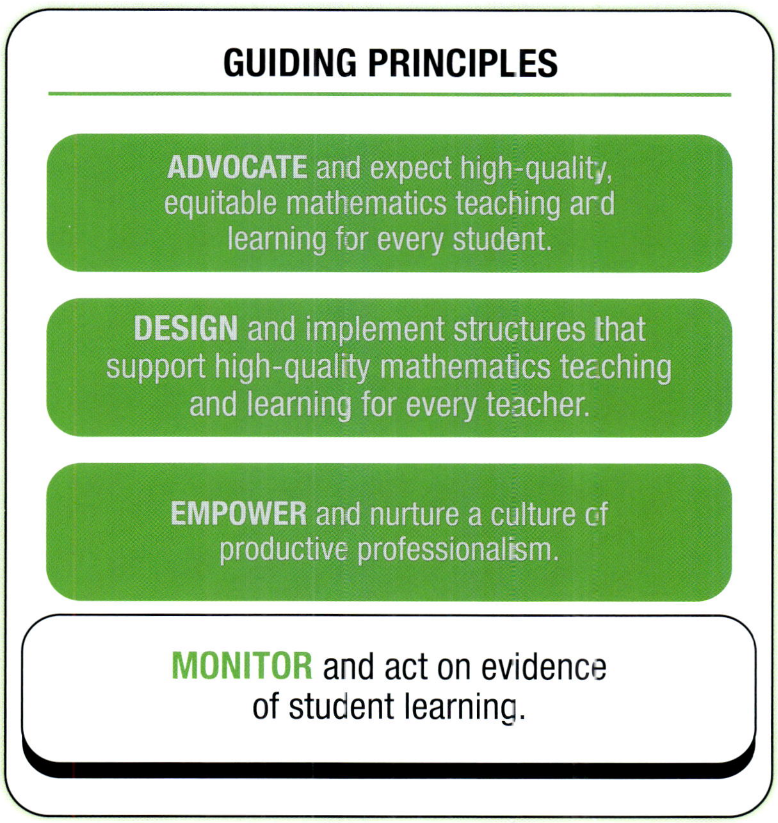

Guiding Principle 4: MONITOR and act on evidence of student learning.

Now that mathematics leaders have **advocated** for students, **designed** structures for teachers to grow their teaching and learning practices, and **empowered** the development of a culture of transparency for learning and reflection, leaders can tackle the fourth and final guiding principle. In classrooms, across teams, and through district programs, the fourth guiding principle for each type of leader is to **monitor and act on evidence of student learning.** Consider the scenarios in Figure 4.1.

Scenario 1: A fourth-grade teacher is teaching a lesson related to addition of fractions with like denominators. Students are working in pairs, completing problems on their white board. Each pair finds another pair to compare their answers and strategies. They use sentence frames like, "We agree with you because…" or "We disagree with you because…" or "Tell us more about this part of your solution." The teacher facilitates the discussion, works to have students correct any misconceptions in their work, and plans the next part of the lesson based on student solutions. Students return to finish their work. At the end of the class period, the teacher gives an exit ticket with three fraction addition with like denominator questions to see which students learned during the lesson. She plans to use any trends in misconceptions as error analysis questions during the start of the next lesson. She will also use the answers to determine how to best continue learning addition of fractions through word problems in the next lesson.
Scenario 2: A high school algebra team is analyzing data from a common end-of-unit assessment related to quadratic functions. They find that students can graph quadratics but have difficulties completing tables for negative x-values and finding zeros when they are not integers. The teachers decide to create stations in their classroom to have some students practice completing tables, others to solve quadratic equations for zeros using factoring or the quadratic formula, and for the students who have demonstrated learning, they will explore solving systems with quadratic and linear equations. Teachers talk about strategies used to teach and realize in one classroom with stronger results, the teacher used an exploration with calculators to graph quadratics and see relationships between a parent function and transformations of the parent function. They determine to try more explorations in the future for new functions students will be learning.
Scenario 3: A new mathematics coordinator is looking at the number of students currently enrolled in grade-level courses to determine how well the district is meeting its mission to ensure the learning of all students at grade level and above. Additionally, she is determining how well students are learning in the courses by looking at state assessment scores and grade data. The results will be analyzed and then used to strengthen the mathematics program. A task force including directors, administrators, and teachers will look at changes in practices, policies, or procedures that need to occur to ensure students are learning grade-level mathematics. They plan to analyze how students are guaranteed access to core instruction and how structures may need to be modified for additional time for teachers to implement intensification targeted lessons, as needed. The task force will also recommend any professional development that may be needed to equip teachers with any skills needed and grow their collective teacher efficacy.

Figure 4.1. Monitor and Act on Student Learning Scenarios

In each scenario above, evidence of student learning was gathered and acted upon. Whether at the classroom level, team level, or district program level, the decisions made using data are at the heart of improving student learning of mathematics.

Teacher leaders, instructional coaches, school-level leaders, district leaders, and state/provincial leaders support the gathering of evidence, use of protocols to make sense of the data trends and patterns, and the instructional and student responses required to ensure student learning. Leaders work to ensure students have access to grade-level content and use evidence from quality assessment processes to monitor student learning and plan next instructional steps, as shown in Figure 4.2, for the last guiding principle, **monitor.**

	Imperative 1 *Commitment to Self*	**Imperative 2** *Commitment to Colleagues*	**Imperative 3** *Commitment to Others*
MONITOR and act on evidence of student learning.	**Essential Action** *Ensure the design and use of high-quality, aligned assessments and equitable assessment processes that guide meaningful reflection and action.*	**Essential Action** *Ensure that the evidence of learning collected from every assessment is used to inform the design of curriculum, instruction, and the assessments themselves.*	**Essential Action** *Ensure every student is provided access to grade-level content and intensification based on evidence of student learning.*

Figure 4.2. Monitor Guiding Principle From the *Framework for Leadership in Mathematics Education*

Within the monitor guiding principle, the commitments to action among the different imperatives emerge as follows.

- Actions related to the first imperative, *commitment to self,* focus on how a leader works with educators in her or his sphere of influence to create and use formative assessments, high-quality common classroom assessments, and high-quality site-based common assessments.
- Actions related to the second imperative, *commitment to colleagues,* focus on how a leader works with educators in adjacent spheres of influence, such as other teacher teams, other departments, on a campus, or within the district, to analyze data from formative assessments and common assessments and to monitor strategic implementation of student intensification plans.
- Actions related to the third imperative, *commitment to others,* focus on how a leader works outside of his or her immediate learning community in order to respond to formative assessment data, create structures for student goal setting, and monitor students' access to grade-level content.

The **monitor** imperatives show the importance of monitoring, and taking subsequent action on, evidence of student learning. In this chapter, we examine these imperatives by focusing on:

- Big Idea 1: Monitoring student learning at the classroom level
- Big Idea 2: Monitoring student learning in teacher teams
- Big Idea 3: Monitoring student learning at the district level

As a mathematics leader, what is your role in monitoring and acting on evidence of student learning? Use these guiding questions in Figure 4.3 to reflect, based on your leadership role, on how you as a bold leader can support the three essential actions within the **Monitor** guiding principle.

Guiding Principle 4: MONITOR and act on evidence of student learning.

Essential Action	Guiding Questions
Ensure the design and use of high-quality, aligned assessments and equitable assessment processes that guide meaningful reflection and action.	In my leadership role: • What do the teachers and leaders I support know about high-quality assessments? • How do we use common scoring, rubrics, and processes for providing feedback? • How do I support teacher teams on my site to develop and use common assessments?
Ensure that the evidence of learning collected from every assessment is used to inform the design of curriculum, instruction, and the assessments themselves.	In my leadership role: • What evidence of learning do we collect and study in collaborative teams? • What structures do we intentionally create to collect, analyze, and act on data from common assessments? • How do we assign students to intervention or intensification programs?
Ensure every student is provided access to grade-level content and intensification based on evidence of student learning.	In my leadership role: • How do we use formative assessment data to provide feedback to students and teachers? • How do we encourage student goal setting? • How do we ensure that all students have access to grade-level content?

Figure 4.3. Leadership Roles With the Monitor Guiding Principle

Visit http://mathedleadership.org/EAresources for a role-specific reflection tool with questions for each mathematics leadership role (teacher leader, coach, site-based leader, district leader, and state/provincial leader).

The imperatives show the importance of using evidence of student learning, as gathered from assessments, to create the instructional and program actions that promote the mathematics learning of every student. Some common types of assessments are shared next.

Assessment Data Types and Leadership Actions

There are several types of data to collect when gathering evidence of student learning. What is the purpose of each? How should each be used to inform program and instructional decisions? Which data most directly impact student reflection and goal setting? Consider the data gathered from each type of assessment in the chart in Figure 4.4 and how they can be effectively used to grow student learning, instructional practices, and mathematics programs.

Mathematics Assessments	Level at Which Data Are Analyzed	Purpose for the Data	Frequency of Data Collection
External Assessments State Assessments Standardized Assessments	• Policy Conversations • State Comparisons • District • Whole School • Department or Grade-Level Team Conversations	• Identify the proportion of students mastering a given set of standards • Evaluate and address effectiveness of systemic processes and structures, instruction, and resources that impact student learning • Determine student population gaps or program gaps in student learning	Once per year
Benchmark Assessments Universal Screeners	• District/System • Whole School • Department or Grade-Level Team • Individual Teacher Analysis	Measures student learning of essential standards at a point in time. Results will inform: • District curricular effectiveness • School instructional effectiveness • District and school professional learning and resource allocations (time, revenue, personnel, etc.) Results are most often not reported as a grade for students but rather to inform program effectiveness and some collaborative conversations about student learning.	No more than three times per year
Common Semester Exams	• District • Department or Grade-Level Team • Individual Teacher Analysis	End-of-semester assessments measure student learning of essential standards at the conclusion of each semester. Teams analyze results to • Determine which students learned essential learning standards • Examine effective instructional practices	Twice per year
Common End-of-Unit Assessments	• Department or Grade-Level Team • Individual Teacher Analysis • Student Analysis	Common end-of-unit assessments measure essential learning standards taught within the unit, including mathematical process standards. Teacher teams develop end-of-unit assessments prior to planning a unit of instruction. Teams analyze results to: • Examine the effectiveness of instruction • Identify students in need of intervention (unless most students need intervention, these data inform Tier 2 interventions) These assessments are graded and should be reassessed when students need additional time and support to master essential learning standards.	At the end of each unit of study during the year (approximately 8–12)
Common Mid-Unit Assessments	• Department or Grade-Level Team • Individual Teacher Analysis • Student Analysis	Developed and analyzed by teacher teams around essential standards. The purpose is to provide information about what students are learning and help decide what to do next when students need additional time, support, and/or enrichment. Teams analyze results to: • Examine the effectiveness of instruction • Determine student learning needs for possible reengagement If graded, treated as a grade that will be adjusted when a student shows improvement based on new learning.	At least one per unit during the year and often two or three depending on the length of the unit
Formative Assessment	• Individual Teacher Analysis • Student Analysis	Created by the classroom teacher to provide teachers and students with real-time feedback about student learning on daily learning targets. May be formal or informal in nature.	Daily

Figure 4.4. Types of Assessment Data to Monitor
Source: Adapted from Finding the Meaning in Numbers (Vagle, 2009).

In addition to assessment data, evidence of student learning is gathered from student work on classwork, homework, or other activities designed to show teachers and students the level of learning acquired. Furthermore, report card grade data may be examined by various stakeholders to measure progress toward a SMART goal, to correlate grades with student learning, and/or as part of placement decisions for students.

How a leader monitors and acts on evidence of student learning gathered through various types of assessments is determined by his or her role and position. Some possible actions are listed in Figure 4.5.

Teacher Leader	• Facilitate discussions related to lesson design that incorporates formative assessment processes with student real-time student feedback. • Orchestrate and facilitate the creation of common assessments and the analysis of data generated. • Use common assessment data to determine effective instructional practices. • Orchestrate team structures for student reflection. • Analyze program effectiveness using large-scale assessment data.
Coach	• Collect and analyze data from multiple sources to inform and improve classroom practices. • Assist the math team leader and/or team to create assessments using protocols for data analysis of student learning. • Support the math team leaders and members with side-by-side coaching focusing on classroom formative assessment practices.
Site Leader	• Provide feedback to teachers, teams, and leaders using data from team and large-scale assessments. • Facilitate the development of a campus-based SMART goal(s) with defined action steps. • Collect and analyze data from multiple sources to inform and improve campus programs.
District Leader	• Facilitate the development of a district goal for effective and equitable mathematics instruction with defined action steps. • Collect and analyze data from multiple sources to inform and improve district programs. • Advocate for students to learn in grade/course-level classes for core instruction in mathematics.
State/Provincial Leader	• Advocate goals for effective and equitable mathematics education policies. • Examine the effectiveness of implementation of statewide or provincewide policies. • Collect and analyze data from multiple sources to inform and improve province or state programs.

Figure 4.5. Mathematics Leadership Actions to Monitor Learning From Assessments

Visit www.mathedleadership.org/EAresources to download a free reproducible version of this figure.

It is important for mathematics leaders to identify the assessment data they need to monitor so the data can best be used to grow teacher and student learning and/or programs and systemic structures and policies. In *Principles to Actions,* NCTM (2014a) states, "Effective assessment of mathematics learning is a process that is coherently aligned with learning goals, makes deliberate use of the data gathered as evidence of learning, and provides guidance for the next instructional steps and programmatic decision making" (p. 92).

Within each school system are several data cycles. The teacher, teacher teams, school, district, and state/province all use data to inform instructional and program decisions. Throughout this chapter, more detailed attention will be given to how mathematics leaders at all levels guide the work of monitoring and acting on evidence of student learning from daily classroom lessons and assessment, teacher team assessments, and large-scale assessment data.

Big Idea 1: Monitoring Student Learning at the Classroom Level

Teachers collect information about students on a minute-by-minute basis. Ranging from a confused look on a student's face during instruction to a well-detailed explanation to their peers, teachers have opportunities to collect "data" on their students' understanding numerous times during the teaching day. How a teacher uses the data supplied by students can determine the focus and appearance of future instruction. The actions of a classroom teacher when monitoring and acting on evidence of student learning fall into the three imperatives shared previously in Figure 4.4. Mathematics leaders work to provide feedback to teachers on their formative assessment processes and stimulate conversations about how to effectively gather evidence related to student learning to inform instructional decisions.

Use and Design of Formative Assessment

Essential Action

Ensure the design and use of high-quality, aligned assessments and equitable assessment processes that guide meaningful reflection and action.

The use of assessment in the classroom takes varied forms. Ultimately, teachers are assessing every time they are checking for understanding during a lesson or giving a more formal exit ticket, quiz, or test. Each one is designed to give information to the teacher that informs next instructional steps, whether in real time during the lesson or in a near-future lesson. NCSM (2014b) states that "effective teachers reflect on student learning and understanding and on their own teaching" (p. 32). So what types of assessments can teachers utilize on a regular basis that would yield such reflections and action? How will a leader support and monitor how the formative assessments are being used for continued student and teacher learning? Such support and monitoring are how effective mathematics leaders honor their *commitment to self* in their work with teachers within their sphere of influence.

NCSM (2014a) defines formative assessment as "a process of gathering evidence within the stream of instruction in order to inform teaching and learning" (p. 1). Two key words within that definition that distinguish formative assessments from summative assessments are *process* and *inform.* Unlike summative assessments, which are generally given at the conclusion of learning (such as a unit test), formative assessments are utilized while the learning is taking place.

Formative assessments could occur during various points in student learning, such as discussions, exploration activities, seat work, homework, and traditional assessments. Another characteristic of formative assessments is that they tell teachers what students need and can guide future instructional sequences. By using the information gathered from students on their current level of understanding and reflecting on the end goal, teachers can design instructional pathways that enhance the learning process for each student. NCSM (2014b) explicitly states, "If teachers do not use the information to adjust instruction, it is not formative" (p. 28).

Formative assessments allow teachers and students to take action during learning, as opposed to after learning, and can help teachers answer questions about their students, such as:

- Which students seem to lack conceptual understanding?
- Which students are on target with grade-level expectations?
- Which students are performing above grade-level expectations?
- Which students are struggling and need additional instruction or an alternate strategy?
- How might a concept need to be differentiated in a current or future lesson?

Formative assessments are similar to a road map or GPS system. By using the current location and proposed destination, the system can create several routes for the journey. Along the journey, as obstacles emerge, the system re-routes and creates new pathways to the original journey. With intentional moments of formative assessments, teachers can do the same for their students.

NCSM (2014b) offers several approaches to implementing effective formative assessment, such as warm-up or bell-ringers, exit slips, questioning, simultaneous response techniques, and student summaries. They can be student work resulting from a question posed to the class or group during instruction. They need to, however, match the target for the lesson or the standard students are learning for the feedback to more effectively reflect what a student has learned or not learned *yet*. Teacher leaders, instructional coaches, and building administrators can work with teams to create such assessments, give feedback to teams regarding the alignment and quality of the assessments, and ask teachers how they plan to learn from the assessment and what they may do instructionally because of it.

Formative assessments are often thought of as a "thing"—a task, written assessment, or activity—that teachers are expected to embed within their instruction. In reality, formative assessments require action on the part of the teacher and the student that takes place after the "thing." When the information from formative assessments is gathered and used to plan the upcoming sequences of instruction and students are receiving validation or making real-time corrections to their thinking and reasoning, the full potential of assessments used for forming instruction is unlocked.

Analyze Data Generated From Formative Assessments

Whether a more formal or informal assessment of student learning, formative assessments provide an opportunity for quick feedback to the teacher and the student regarding student learning. A teacher can intentionally design an assessment to gather information regarding such varied aspects of student learning related to a concept to include each student's

- understanding (including misconceptions),
- ability to apply any needed prerequisite skills,
- conceptual understanding,
- procedural understanding,
- confidence in his or her own ability to be successful, or
- choice of strategy or representation to tackle mathematics tasks.

Leaders, honoring their commitment to colleagues, work with teachers to identify the purpose of each formative assessment and provide structures for the teacher to analyze the information gleaned from each one.

Essential Action

Ensure that the evidence of learning collected from every assessment is used to inform the design of curriculum, instruction, and the assessments themselves.

For example, Figure 4.6 shares a template teachers might use to record their student learning data as they think about an instructional response to a formative assessment made to a target or standard.

Target or Standard:			
	Proficient	Close to Proficient	Far From Proficient
Number of students or Names of students in each classification.			
Patterns or trends in student work: What in their work is the evidence of how their learning should be classified?			
How will each group of students reengage in learning?			

Figure 4.6. Data Analysis Protocol for a Classroom Formative Assessment

Visit www.mathedleadership.org/EAresources to download a free reproducible version of this figure.

Mathematics content standards often comprise several components that collectively describe student mastery. Through the use of formative assessments, teachers can gauge students' development and mastery of the subcomponents prior to evaluating student mastery of the standard in its entirety. One advantage to gauging student mastery along the way is the ability to immediately act when additional learning is needed.

In the classroom, a teacher might see an error in student work and work with the student to make the correction. After giving a more formal exit ticket at the end of the lesson, the teacher can determine any gaps in student understanding by looking at trends in student responses to see who has learned and who needs additional instruction. Leaders can observe this in action through walkthroughs, discuss strategies for incorporating formative assessment into lessons, and ask teachers how they know students learned or did not learn during a lesson.

Additional data from formative assessments can be collected related to student confidence. One way to gauge student confidence in learning is to incorporate response items that allow students to share their level of confidence with the content being learned, such as:

- How did you feel learning the material?
- How did you feel while completing this assessment?
- What was the hardest part of this material for you?
- Which describes your confidence: (1) Do you feel you can do this target with help? (2) Do you feel you can do this target on your own? Or (3) Do you feel you can teach others this target?

Because an assessment is only formative if teachers and students take action on the learning, consider how students are learning from each formative assessment. As leaders, what supports or learning might teachers need to have students revise their thinking and set goals? For less formal formative assessments that happen during a lesson when students receive feedback related to corrections in their thinking or reasoning, how are students taking action and fixing misconceptions early in their learning? Leaders will provide the structural time in the day for lessons to incorporate the time needed for tasks that illicit these formative assessment moments. Leaders also work with teachers to learn strategies for effective feedback and monitor and support teachers working to provide daily feedback to students.

Once the data have been analyzed by a teacher, formally or during a lesson, how should the teacher and student respond to the feedback? The next section explores some ideas related to differentiation and intensification.

Respond to Formative Assessment Data

Essential Action

Ensure every student is provided access to grade-level content and intensification based on evidence of student learning.

What happens when it is clear that a student has a misconception during a lesson? Does a teacher simply walk away? Of course not! Teachers work to ensure the learning of every student so they will think about how to scaffold learning up to the level of the task. What manipulatives might need to be on hand for students to make sense of the task? What questions might the teacher ask to give students access to the task? How might the teacher also extend the task if students finish early (Kanold et al., 2018)? During a lesson, answering such questions allows a teacher to differentiate instruction and engage all students in learning mathematics. Leaders, honoring their *commitment to others,* work with teachers to ensure they have the resources and manipulatives needed to meet the needs of each student and provide feedback to the resulting student learning from the differentiation.

From a more formal formative assessment, leaders work with teachers to design strategies and structures for continued learning with students. Leaders challenge teachers to plan for or work with teachers to find time to reengage students in learning.

Just as importantly, mathematics leaders work with teachers to reflect on their own instructional practices. After finding that a majority of the class did not understand a topic, a teacher's reflection may prompt an investigation into his or her own content knowledge and instructional practices. The teacher may realize that student mastery of the content may hinge on seeking specific content or pedagogy-centered professional development. In this sense, the teacher is using the students' formative assessments as a formative assessment of their own instruction!

Developing a habit of formatively assessing begins with purposeful planning. Teachers need to begin by establishing goals for each lesson and the instructional activities associated with them. Student learning expectations should be clear to all stakeholders, including teachers, and students. NCSM (2014b) states that the content should be grade-level appropriate and clearly articulated to students. Formative assessment is not just a portion of lesson plans or administrators' expectations but instead is a thought-out, well-designed series of checkpoints that inform teachers on what the next lesson should look like for students.

Students also respond to their formative assessment data so they are part of the full learning cycle. Teachers and students work together to develop a plan of forward progression to the current goals and detail the work it will take to get there. These conversations with students can take many forms. One suggestion from NCSM (2014a) is to allow students to analyze anonymous student work samples to communicate the strengths and weaknesses of the samples in regard to the learning goal. These conversations give students meaning to the work they are doing and can provide examples of the end expectations. Another suggestion is to provide students with learning rubrics or descriptors of levels of learning. These tools can be beneficial before, during, and after learning to guide students in understanding and reflecting on expectations. When used prior to learning, students can begin by establishing a clear understanding of the goal and the measurements for mastery. During learning, students can gauge their development toward mastery, and after learning, students can reflect on their progress.

> **"When done correctly, feedback can result in students reflecting and rethinking their mathematics, while increasing their effort and motivation."**
>
> **—NCSM (2014a), (p. 2).**

Providing feedback to students allows them to reflect, and act on, their own learning. According to NCSM (2014a), "When done correctly, feedback can result in students reflecting and rethinking their mathematics, while increasing their effort and motivation. Providing feedback has been shown to have positive effects on student learning" (p. 2). The goal of providing students with feedback is to give them an avenue to take action on their learning. Leaders work with teachers to create the culture and structures necessary for meaningful student analysis of learning.

Monitoring student learning in the classroom involves mathematics leaders working with teachers to implement formative assessment processes that allow teachers and students to take action based on the evidence of student learning, both formal and informal. Leaders identify effective practices leading to increased student learning and the professional development that may be required to better equip teachers to routinely check and respond to student learning.

While it is important for teachers in the classroom to monitor and act on evidence of student learning, mathematics leaders also work with and guide teacher teams working collaboratively to determine what students must learn and responding to that learning for equity purposes as a collective team.

Big Idea 2: Monitoring Student Learning in Teacher Teams

From the state/provincial level to the teacher level, teams collect evidence of student learning using quantitative and qualitative data. State assessments, benchmark assessments, and classroom assessments provide numerical data related to student learning. Evidence of learning using student work or classroom observations provides an opportunity to learn about student reasoning using more qualitative measures focused on finding trends in student thinking for a targeted instructional response, as needed. While assessments can be given and evaluated independently across a grade-level or course-alike team, a commitment to colleagues requires teachers to learn from one another using some common assessment data aligned to what the team expects all students to know and be able to do.

When teachers work in isolation from one another, inequities in student learning are too often created. The standards or level of complexity a student is expected to learn vary from one classroom to the next. DuFour (2016) states, "A genuine commitment to helping all students learn at high levels requires a systematic process to monitor each student's learning on an ongoing basis and then use evidence of student learning to:

- Provide students with relevant, actionable information about the progression of their learning towards clearly defined standards and targets.
- Respond to the individual needs of students in a coordinated way.
- Inform and improve the individual and collective practices of professionals within the school." (p. 157)

Mathematics leaders monitor the effectiveness of grade-level or course-alike practices related to common assessments. In other words, how are teacher teams creating high-quality assessments and then using the assessments to effectively inform instructional practices, analyze student learning to determine which students learned and possible intensification needs, and provide an opportunity for student reflection and goal setting? The creation and use of such common assessments are critical to establishing a genuine guaranteed and viable curriculum for student learning.

Create High-Quality Assessments

Essential Action

Ensure the design and use of high-quality, aligned assessments and equitable assessment processes that guide meaningful reflection and action.

As a leader of mathematics education, it is important to first monitor the creation of quality assessments aligned to the standards in a unit if it is to be used effectively for continued teacher and student learning. Doing so honors the leader's *commitment to self* as this is done in concert with teachers and leaders in her or his immediate sphere of influence. When created before a unit of study, teachers clarify the standards or learning targets students will learn in the unit as well as the rigor of the tasks required to demonstrate proficiency of each standard, which better informs the decision making needed for lesson design throughout the unit.

The Assessment Instrument Quality Evaluation Tool in Figure 4.7 (Kanold et al., 2018) shows eight critical attributes of high-quality assessments in a single rubric. Teams and leaders can use this tool to determine strengths and weaknesses in current assessment design practices for mathematics grade-level and course-alike teams. Next, teams can use the feedback to grow the effectiveness of their assessment instruments.

Assessment Indicators	Description of Level 1	Requirements of the Indicator Are Not Present	Limited Requirements of This Indicator Are Present	Substantially Meets the Requirements of the Indicator	Fully Achieves the Requirements of the Indicator	Description of Level 4
Identification and emphasis on learning targets (specific feedback to students)	Essential learning standards are unclear, absent from the assessment instrument, or both. Some of the mathematical tasks (questions) may not align to the essential learning standards of the unit. The organization of assessment tasks is not clear.	1	2	3	4	Essential learning standards are clear, included on the assessment, and connected to the assessment tasks (questions).
Balance of higher-and-lower cognitive demand mathematical tasks	Emphasis is on procedural knowledge with minimal higher-level-cognitive-demand mathematical tasks for demonstration of understanding.	1	2	3	4	Test is rigor balanced with higher-level and lower-level-cognitive-demand mathematical tasks present and aligned to the essential learning standards.
Variety of assessment task formats and use of technology	Assessment contains only one type of questioning strategy—selected response or constructed response. There is little to no modeling of mathematics or use of tools. Use of technology (such as calculators) is not clear.	1	2	3	4	Assessment includes a blend of assessment types and modeling tasks or use of tools. Use of technology (such as calculators) is clear.
Appropriate and clear scoring rubric (points assigned or proficiency scales)	Scoring rubric is not evident or inappropriate for the assessment tasks.	1	2	3	4	Scoring rubric is clearly stated and appropriate for each mathematical task.
Clarity of directions	Directions are missing and unclear. Directions are confusing for students.	1	2	3	4	Directions are appropriate and clear.
Academic language	Wording is vague or misleading. Academic language (vocabulary and notation) is not precise, causing a struggle for student understanding and access.	1	2	3	4	Academic language (vocabulary and notation) in tasks is direct, fair, accessible, and clearly understood by students. Teachers expect students to attend to precision in response.
Visual presentation	Assessment instrument is sloppy, disorganized, difficult to read, and offers no room for student work.	1	2	3	4	Assessment is neat, organized, easy to read, and well spaced, with room for student work. There is also room for teacher feedback.
Time allotment	Few students can complete the assessment in the time allowed.	1	2	3	4	Test can be successfully completed in the time allowed.

Figure 4.7. Assessment Instrument Quality Design Tool
Source: Kanold et al. (2018). Used with permission. From *Mathematics Assessment and Intervention in a PLC at Work*™ by Timothy D. Kanold, Sarah Schuhl, Matthew R. Larson, Bill Barnes, Jessica Kanold-McIntyre, and Mona Toncheff.

Because teachers will use data from assessments to give feedback to students and determine instructional next steps, it is important that the fourth attribute of quality assessment is addressed when the assessment is created. How will the assessment be scored? Consider how teams determine and document their common scoring agreements when designing each assessment, whether using points for items on the assessment or a proficiency scale for standards on the assessment. Knowing student scores are aligned allows for meaningful data dialogue discussions and more accurate intensification plans, if needed.

Creating and using high-quality assessments is an important step in monitoring evidence of student learning. We must also act on the evidence of learning generated by those assessments. In the next section, we will explore ways to analyze that evidence.

Analyze Data From Assessments

Essential Action

Ensure that the evidence of learning collected from every assessment is used to inform the design of c urriculum, instruction, and the assessments themselves.

Data analysis is at the heart of this essential action monitoring how assessment evidence is used to inform the design of curriculum, instruction, and the assessments themselves. Consistent use of data analysis protocols and techniques allows mathematics leaders to honor their *commitment to colleagues* as the consistency expands into the leader's adjacent spheres of influence. Protocols allow colleagues to maintain a focus on the analysis of student learning with action steps while building trust, a required element in the process.

Team Analysis of Large-Scale Assessment Data

Large-scale assessments provide insights into strengths and weaknesses of programs and student learning but seldom provide targeted information by standard due to the nature of the assessments and the minimum number of questions that can be asked for any one standard. Leaders at all levels can use the data from large-scale assessments to strengthen the learning of content and process standards and analyze the level of equity in learning between various student populations.

Some questions that might be used to analyze the data are in Figure 4.8, using an adapted *Here's What, So What, Now What* protocol from Wellman and Lipton (2004). This protocol is designed to be completed with colleagues for a team response, whether a district, school, or grade-level team.

Here's What, So What, Now What Data Analysis Protocol

Analyze the data from the mathematics assessment for the entire grade level or course to address each part of the protocol.

Here's What

What are the data the team needs to be analyzing? Simply recognize the data, categories, and student populations. This is not yet the time to critique or make inferences about student learning in the data.

- Which aspects of mathematics should be examined (e.g., strands, standards, mathematical process standards)?
- Which student populations need to be examined for equity in student learning?
- If comparisons are to be made, should they be made looking at students as a cohort from one grade or course to the next or as part of a program looking at how students in the grade level or program did last year on a similar assessment to the students in the same grade level or course this year?

So What

Recognize trends from the data without yet making inferences or an action plan. Consider using the sentence frame, "I notice..." when articulating observations.

- What do you notice in the data?
- Which content is a strength for students in the grade level or course?
- Which content is a weakness for students in the grade level or course?

Now What

Make conclusions and inferences about the data to structure a collective response to student learning.

- Why might students have scored well or not scored well in light of instructional practices or programs? What will we do about it this year? Next year?
- How do the results show equity or inequity in learning between various student populations? Why? How will we address any inequities this year? Next year?

Conclude with an action plan and clear role responsibilities for each person involved in analyzing the data.

Figure 4.8. Here's What, So What, Now What Data Analysis Protocol
Source: Adapted from Wellman and Lipton (2004).

Visit www.mathedleadership.org/EAresources to download a free reproducible version of this figure.

While large-scale assessments do provide information relevant to teacher teams, teachers often need more specific information than the assessments can provide in order to most effectively design quality instruction, identify intensification needs, and celebrate learning successes. Such information results from analysis of common assessment data.

Team Analysis of Common Assessment Data

Teachers work to have students learn mathematics standards in an ongoing cycle of planning, instruction, assessment, reflection, and action for each mathematics unit of study. Each cycle of study begins by first clarifying the standards students must learn and therefore what students must know and be able to do. Team common assessments are then created and their data used as part of a data dialogue to determine which students have learned or not learned yet, to identify effective instructional practices, and to design quality reengagement lessons as needed.

Throughout each unit of study, both common summative (used to evaluate student learning and programs) and formative assessments (used to inform instructional decisions and monitor student progress) are used. Teams work to balance the implementation and use of summative and formative assessments (Stiggins, 2014). NCSM (2008) reminds us that "our understanding of assessment must change from seeing it as mere after-the-fact testing to using it as an integral component of the planning-teaching-assessing cycle that characterizes high-quality instruction" (p. 46). In other words, while each type of assessment may serve a unique purpose, any classroom assessment can be used formatively for teacher and student learning of mathematics standards when students have not yet learned.

Mathematics leaders and teachers determine how to effectively use assessment data for continued teacher and student learning. They determine the trends in student reasoning and identify which standards or learning targets students might need additional time and support to learn. Leaders of mathematics work to allocate time for teacher teams to effectively plan a response to common assessment data. They also support teams as they gather and organize evidence of student learning and use protocols to determine which students have learned each standard or learning target and which instructional practices proved most effective for student learning.

One protocol that teams might use to monitor student progress on a common mid-unit or formative assessment is shown in Figure 4.9. This protocol would be completed for each essential standard and/or learning target assessed.

Unit:		**Team:**	**Date:**	
Standard and/or Learning Target:				
	Number of Students	**Trend(s) in Student Work** *(What strategies did students use? What do they understand?)*	**Effective Instructional Practices** *(Which instructional practices contributed to student learning?)*	**Instructional Plan** *(What is a team reengagement plan to continue student learning?)*
Students demonstrating proficiency				
Students demonstrating a partial understanding				
Students showing a minimal understanding				
No evidence of student learning				

Figure 4.9. Team Common Mid-Unit Data Analysis Protocol

Visit www.mathedleadership.org/EAresources to download a free reproducible version of this figure.

Figure 4.10 shows a protocol for a shorter common assessment. Although it can be used for larger common end-of-unit analysis of data, mathematics leaders may also want to consider the following protocol for use with teams. Notice that regardless of protocol used, teams need to analyze student learning by standard and/or learning target.

Percentage of Students Demonstrating Proficiency or Above				
	Standard/Target 1	Standard/Target 2	Standard/Target 3	Standard/Target 4
Teacher A				
Teacher B				
Teacher C				
Team Total				
Use student work to complete the protocol.				
For each standard/target, what are trends in the work of students at each proficiency level?				
	Thorough Understanding	Adequate Understanding	Partial Understanding	Minimal Understanding
Standard/ Target 1				
Standard/ Target 2				
Standard/ Target 3				
Standard/ Target 4				
Which instructional practices or mathematical strategies contributed to student learning?				
What is a possible team response for intensification or extension of student learning for each level of proficiency?				
	Thorough Understanding	Adequate Understanding	Partial Understanding	Minimal Understanding
Standard/ Target 1				
Standard/ Target 2				
Standard/ Target 3				
Standard/ Target 4				
What changes, if any, to the assessment are needed to generate more accurate or meaningful data?				
From the data, what can our team celebrate related to student learning as a result of our instruction?				

Figure 4.10. Team Common End-of-Unit Data Analysis Protocol

Visit www.mathedleadership.org/EAresources to download a free reproducible version of this figure.

While analyzing data, teams may also need to consider how various student groups performed to more effectively create a team response to learning. Additional protocols teams might use can be found at www.mathedleadership.org/EAresources.

NCSM (2008) states that not just the analysis of data matters, but rather the interpretation and action that follow to grow student learning. "The leader must understand the importance of gathering and interpreting relevant assessment data that contribute meaningful feedback for students, teachers, and administrators—that contribute meaningful feedback for students, teachers, and administrators—all in the service of improving student achievement for each and every student" (p. 46). Leaders work to ensure that the protocols shared do not end with identification of students who learned or did not learn, but rather with a plan to continue the learning of each student group identified by levels of proficiency.

Respond to Evidence of Learning on Assessments

The data from quality common assessments allow a team to make data-informed intensification decisions as a strategic and targeted response to student learning. Teachers determine which students need additional time and support to learn standards and create a plan to reengage students in learning. Teams may need to consider how to enrich or extend learning when addressing learning gaps during an intensification plan. The plans for such intervention or extension were a part of the team assessment protocols in Figures 4.7 and 4.8.

Mathematics leaders monitor whether or not teams are identifying trends in student learning and determining an instructional response as part of an intensification process. Additionally, they monitor whether or not teams are determining the effectiveness of any intensification used and how students are placed based on their learning of standards rather than only external assessment data or universal screener data.

When monitoring the intensification plan of teacher teams, consider the questions shown in Figure 4.11.

Questions to Monitor Team Intensification Plans

1. How did your common assessment data drive your team intensification plans for students?
2. When will your team implement your intensification plan for students?
3. How will students be put into the appropriate reengagement activity during your team intensification plan?
4. How will your team know whether or not student learning improved at all levels as a result of your intensification plan?

Figure 4.11. Questions to Monitor Team Intensification Plans

Visit www.mathedleadership.org/EAresources to download a free reproducible version of this figure.

Not only do teachers respond to the evidence of student learning, but students also are an integral part of the process. How will students know whether or not they have learned or are learning each standard or target?

Create Structures for Student Reflection and Goal Setting

Collective team action using common assessment data does not stop with a data analysis protocol and intensification plan. A team's strong use of grade-level common assessment data involves using structures to promote students' self-reflection of learning related to each standard or target. Bringing students, as stakeholders in their own learning, into the process honors the leader's *commitment to others.* Stiggins, Arter, Chappuis, and Chappuis (2006); Marzano (2006); and Hattie, Fisher, and Frey (2017) provide evidence of gains in student learning when students reflect and learn from their mistakes, graph their learning for each standard, and set goals for continued learning.

Essential Action

Ensure every student is provided access to grade-level content and intensification based on evidence of student learning.

In elementary school, teams may start by placing three faces at the bottom of a common assessment and have students circle which one best describes how they feel they learned the standard or target. A teacher can then circle the reflection as well and then have conversations with students when there are any discrepancies. See Figure 4.12.

Mathematics leaders look for student reflection and feedback structures used across a team for equitable student learning experiences. They work with teams to effectively use the structures with students for learning.

Student reflection tools work best when students have been introduced to the standards or learning targets at the beginning of the unit and they are used in every lesson and practice task. As such, at the end of a unit, another reflection tool might look like the one in Figure 4.13.

Figure 4.12. Student Reflection Example for Common Assessments

Unit: Systems of Equations	Name:	Date:
Identify your score for each learning target.		
Learning Target	**Score**	**Circle One**
I can graph a system of linear equations.		Got It! Still Learning It
I can solve a system of linear equations and explain my reasoning.		Got It! Still Learning It
I can determine when a system of linear equations has one solution, no solutions, or infinitely many solutions.		Got It! Still Learning It
Answer each question. What have I learned? What do I still need to learn? What is my plan to still learn?		

Figure 4.13. Grade 8 Student Reflection Example
Source: Adapted from Kanold et al. (2018).

Visit www.mathedleadership.org/EAresources to download a free reproducible version of this figure.

Leading and monitoring the creation of quality common assessments and effective data analysis for teacher teams is an essential action for mathematics leaders at every level. When teams engage in creating the common assessments before each unit begins as part of the teaching-assessing-learning cycle, their daily lesson design and feedback to students strengthen. Working as a team to make an intensification plan for all students across a grade-level or course-alike team works to create equity in learning as teachers work together to have students learn each standard or target.

Big Idea 3: Monitoring Student Learning at the District Level

NCSM's (2008) *PRIME Leadership Framework* challenges mathematics leaders to continuously work to "seek out and erase biases and inequities that exist in student learning and assessment experiences" (p. 10). This work requires a bold commitment by leaders of mathematics to regularly analyze and interpret large amounts of grade and assessment data in their efforts to ensure that every student is provided access to grade-level content and intensification based on evidence of student learning. It requires organizational leaders to look honestly at the system, identify gaps in equity, inspect access and performance, and communicate and lead teams in next steps and actions intended to move the system into more equitable and excellent outcomes.

Monitoring these data is not the work of a single year, but the work of identifying trends in student learning over time as leaders seek to ensure that outcomes are truly becoming more equitable. The educational systems in schools and districts are complex with a series of interconnected relationships and interactions. Understanding these connections and relationships takes time and diligence. As Aguilar (2013) states in *The Art of Coaching,* "We must seek to understand these interactions in order to intervene effectively to change them" (p. 54).

Mathematics leaders demonstrating a commitment to others monitor the access students have to quality instruction, grade-level content, and effective intensification annually, as well as work to mobilize teams of educators to understand and respond to the data so that over time, the learning outcomes of the institution become more equitable. Leaders need to inspect and answer these important questions:

- Are traditionally underserved populations in the school having more or less success?
- How are students fairing with grade-level content?
- Are all students getting access to grade-level content?
- Are intensification plans improving student learning outcomes over time?
- What are all the possible factors that are contributing to the student learning outcomes?

The ongoing and courageous work of monitoring the outcomes of a complex instructional system is critical to establishing "a systematic plan for the continuous improvement of student achievement across all populations" (NCSM, 2008, p. 12). At the district level, mathematics leaders may look more strategically at common district or state assessments, implementation of intensification plans, and how students are getting access to and learning grade-level mathematics.

Common Site-Based Assessments

Essential Action

Ensure the design and use of high-quality, aligned assessments and equitable assessment processes that guide meaningful reflection and action.

If assessments are to be used to effectively monitor and act on evidence of student learning, then that evidence must be collected in ways that are mindful of equitable assessment practices. Leaders, while honoring their *commitment to self* while working in their spheres of influence, must be mindful of the potential inequities that can be induced by an assessment system. Hence, "of utmost importance is the *imperative* for adult *collaboration* in the development of assessments and the reflection on the results of assessment" (NCSM, 2009, p. 1).

As we explored earlier in the chapter, at minimum, the assessments must be "congruent and aligned by grade level or course content" (NCSM, 2008, p. 49). One way to guard against potential inequities among teachers and course sections is for educators to collaboratively develop common assessments that all students enrolled in a particular grade or course take.

Common assessments are developed through a process (Figure 4.14) that establishes *what* students are to learn, *how* their learning will be assessed, and *what* the evidence of learning reveals. The evidence of learning becomes data that can then be used to inform revision or refinement of the assessment itself. The data can also be used to determine the effectiveness of curriculum or instruction.

Systems of high-quality assessment help districts, schools, teams, and teachers address the question, "How will we know students are learning what is intended?" Effective leaders of mathematics create systems that support the development of common assessments at the district level, at the school level, and in the classroom. We consider assessment to be a multifaceted process by which we gather information about students, teachers, schools, and school districts to provide formative opportunities to learn, adjust instruction and lesson planning, inform decision making, and monitor progress and effectiveness. In order to promote and support changes in classroom practices that ensure all students are learning, teachers must balance *assessment for learning* (used to form instructional decisions and monitor student progress) with *assessment of learning* (used to evaluate students' progress and achievement, assign grades, and appraise programs) (Wiggins & McTighe, 2005).

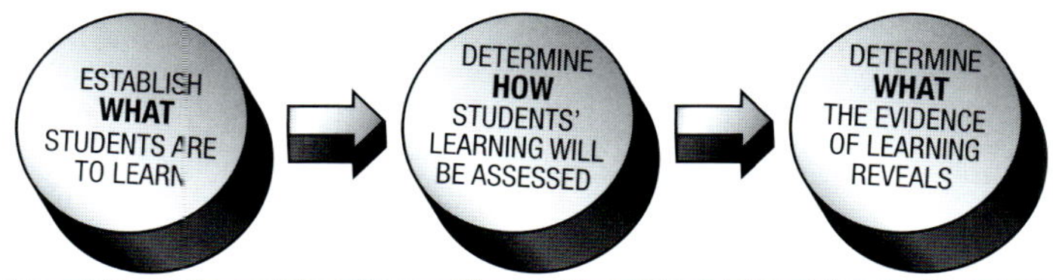

Figure 4.14. Process for Developing Common Assessments

Of utmost importance is the imperative for adult collaboration in the development of assessments and the reflection on the results of assessment. For far too long, assessment practices have been a primary creator of inequity sorting out students due to the isolated decision making and inconsistent learning expectations. When school mathematics education leaders engage teacher teams in a collaborative process that establishes what students are to learn, how their learning will be assessed, and what the evidence of learning reveals, they promote the clarity and consistency in purpose and implementation that are necessary to reduce bias.

Equitable student learning experiences can only occur when teacher collaboration trumps the student assessment inequities created by teacher isolation alongside failure to plan, teach, and assess using research-informed best assessment practices. "How will we know our students are learning?" must become a leadership question that is answered in the collaborative.

Mathematics leaders must act to recognize the importance and impact of the skillful use of assessment on teacher practice and student learning. Highly effective and collaboratively developed assessments will create greater equity through consistency in rigor as student assessments are aligned with curriculum and instructional planning.

Figure 4.15 highlights leadership actions forwarded in the 2009 NCSM position paper, *Improving Student Achievement in Mathematics by Leading Highly Effective Assessment Practices.*

More specifically, effective leaders of mathematics must engage teachers, teacher teams, and teacher leaders in collaborative discussions and actions to:	
1.	Increase the use and types of assessments for learning to build student confidence in themselves as learners.
2.	Focus on the coherency in and consistency of the grading criteria and accountability used for individual and team assessment purposes.
3.	Implement assessment practices and policies as a means of instructional planning and improvement. This includes but is not limited to helping teachers become competent masters of the standards students are to master, deconstructing each standard into achievement targets that help students master the standard and sharing the targets with students from the beginning of the learning using clearly defined terms and accessible language to create high-quality assessments that reflect those targets, and using those assessments (in collaboration with other teachers) with students to track improvement over time (Stiggins, 2007).
4.	Use formative assessment process on data from summative assessments to improve student learning through ongoing student feedback regarding criteria for success.
5.	Translate classroom assessment results into frequent, descriptive (versus judgmental) feedback for students, providing them with specific insights regarding their strengths as well as how to improve (NCSM, 2008).
6.	Adjust instruction continuously based on the results of formative and summative assessments (NCSM, 2008).
7.	Evaluate local assessment quality and ensure alignment with state and national curriculum recommendations.
8.	Utilize local district benchmarks to provide formative feedback loops to the teacher team as well as to the students.
9.	Use district, state, or provincial summative data as an integral part of the total analysis of student learning and mathematics program decisions.

Figure 4.15. Assessment Leadership Actions

Visit www.mathedleadership.org/EAresources to download a free reproducible version of this figure.

At the campus or district level, common assessments frequently take the form of performance assessments or benchmark assessments that are standards based and contain items that measure student understanding of particular standards or learning goals. Data from these common assessments are then used to answer several questions, including the following:

- How well are our teachers implementing the intended curriculum?
- How well are all student populations performing when compared to their peers?
- Which students need intensification plans?

Leaders can analyze data using a method shown in Figure 4.16 to determine how well students have learned the intended curriculum.

Standard	Description	<50% Correct	50%–75% Correct	>75% Correct

Figure 4.16. Standard Analysis for Common Assessments

Mathematics leaders can reflect on the results to identify possible next steps. For example, if less than 50% of students answered questions about a standard correctly, then there is likely a misalignment between the assessed curriculum and the instruction that students received. In this case, district leaders can work with the teachers they serve to determine whether the learning standards were taught prior to the assessment and, if so, how they were taught.

If between 50% and 75% of students answered questions about a standard correctly, then it is likely that the concepts and skills were taught in alignment with the assessed curriculum. Rather, students did not know the concepts and skills well enough to demonstrate mastery. In this case, district leaders can help teachers or teams revisit how they taught particular concepts and skills and develop strategies for reinforcing them.

If more than 75% of students answered questions about a standard correctly, then it is likely that most students have mastered this content. District leaders can help teachers and teams develop strategies to maintain these concepts and skills for students who have mastered them and develop intensification or reteaching strategies for students who have not.

Essential Action

Ensure that the evidence of learning collected from every assessment is used to inform the design of curriculum, instruction, and the assessments themselves.

Monitor the Strategic Implementation of Intensification Plans

An important element of monitoring and acting on evidence of student learning is to monitor the strategic implementation of intensification plans. Mathematics leaders honor their *commitment to colleagues* while working alongside other spheres of influence for consistency in how the plans are implemented. Raising achievement in mathematics for every student in every classroom requires "a variety of *intensification* strategies be available to teachers to support the learning needs of struggling students" (NCSM, 2014b, p. 35). The use of the term *intensification* rather than *intervention* is purposeful.

Rather than responding with an intervention strategy *after* students have failed, successful schools take a proactive stance by anticipating the learning needs of students that are not met within regular 50- or 60-minute class periods and by implementing strategies that *intensify* the learning experience (NCSM, 2014b, p. 35).

By its nature, intensification is not a one-size-fits-all approach. Leaders and teams of leaders must ensure that multiple intensification strategies are available to support the range of learning needs of struggling students.

- Consider and employ research-based support structures that will best fit the needs of the student population. Successful strategies move students forward by expanding the learning experience, including offering daily outside-the-classroom help, extending class time, and offering bridge or preparatory courses.
- Identify students who need additional support, utilizing multiple data sources. These data sources may include data already collected for course placement purposes; input from teachers, parents, and students; state assessment results; or benchmark or interim assessment results.
- Ensure access to intensified learning opportunities for identified students. Schools should have a comprehensive constellation of support structures based on the anticipated learning needs of struggling students.

In this section, we will examine ways to respond to evidence of learning on assessments so that students have equitable access to high-quality instruction as well as the opportunity to fill in gaps in their prior understanding of mathematics content.

Respond to Evidence of Learning on Assessments

Earlier in this chapter, we discussed how teacher teams can use common assessments and the data they generate to inform intensification strategies. At the district level, mathematics leaders may use similar tools to analyze common assessment data. A difference is that district-level leaders look through a different lens as their responsibilities regarding intensification strategies focus on making sure that individual campuses and teachers in their district have the resources, tools, and skills they need in order to effectively implement intensification strategies.

When monitoring the implementation of intensification plans, consider the questions shown in Figure 4.17.

Questions to Monitor Implementation of Intensification Plans

1. How did your common assessment data drive your team intensification plans for students?
2. Which intensification strategies did the team select for which students?
3. What student performance data did the team select to monitor the intensification strategy(-ies)?
4. How will your team know whether or not student learning improved at all levels as a result of your intensification plan?
5. What additional resources and tools do teachers need to better implement the intensification plan?

Figure 4.17. Questions to Monitor Team Intensification Plans

Visit www.mathedleadership.org/EAresources to download a free reproducible version of this figure.

Reflective questions like these help leaders identify trends in data as they seek to answer the main question, "Was this intensification strategy effective?" As with any classroom instruction effort, it is important to know during the intensification if it is being effective, so leaders will want to gather and analyze interim data, such as student work samples, test or quiz results, and so on, to determine whether or not the intensification program is on target to meet its particular goals for that particular group of students.

When analyzing data to evaluate the effectiveness of programs, there is a tendency among team members to rush to judgment about why a particular element of the program is or is not working as intended. Protocols and discussion tools, such as the Data Analysis Protocol in Figure 4.18, can be used to intentionally slow down the discussion and allow multiple ideas to emerge and be considered.

Exploratory Dialogue: Activating and Engaging

(Prediction: What do you predict the data and results will show?)

Exploring and Discovering

(Please avoid because statements or trying to explain WHY the data are what they are…we will do that part next. Just observe and note the patterns, trends, etc. that are emerging.)

Exploratory Dialogue: Organizing

What are possible causes for these data and results?

STUDENT knowledge, skills, and dispositions	**INFRASTRUCTURE** schedules, programming, and resources	**CURRICULUM** design and implementation	**INSTRUCTION** methods, materials, and resources	**TEACHERS** knowledge, skills, and dispositions

Wellman's Theories of Causation

Theory 1

Theory 2

Theory 3

Exploratory Dialogue: Brainstorming Solutions

(What can we do individually and as a team to address the data issues that we noted above? Keep in mind the theory of causation and how your solution will address the data and results.)

Decision Making and Next Steps

(Create an action plan for next steps. What actions will you and your team take? What is the timeline? What criteria will you use to determine success?)

Figure 4.18. Data Analysis Protocol

Visit www.mathedleadership.org/EAresources to download a free reproducible version of this figure.

In this data protocol, participants are first asked to individually make predictions about the data before being viewing the data. Groups then share their predictions with each other as well as "why" they made those predictions. Only after this step is the group presented with the data set. Participants then work through the remaining parts of the protocol, taking care to not rush too quickly to decision and action steps.

As you analyze the effectiveness of the intensification plan, it is important to include reflection on how well the program is serving all students. Intensification involves identifying some students for additional instructional support beyond their regular classroom instruction. Any time groups of students are targeted for particular activities, there is a potential for introducing inequities into the system. Effective leaders are mindful of this potential and actively guard against it, including monitoring which types of students (ethnic group, socioeconomic status, gender, English language fluency, etc.) are being included in the intervention groups. If the student population in an intensification group isn't representative of the student population as a whole, mathematics leaders question why that is and seek ways to rectify the inequity.

Monitor Access to Grade-Level Content

Access to grade-level content is at the core of the essential action in a mathematics education leader's *commitment to others.* In the work to deliver equitable and high-level learning outcomes, monitoring a school or district's ability to help students succeed at grade level must be a regular and ongoing practice. Mathematics education leaders understand that "no intervention program can compensate for ineffective core instructional practices" (Buffum, Mattos, & Weber, 2009, p. 74) and must work to monitor, adjust, and continuously improve their core instructional program based on evidence of student learning.

Essential Action

Ensure every student is provided access to grade-level content and intensification based on evidence of student learning.

It might be necessary to conduct a *Where Are They Now* student access survey (Figure 4.19) to provide an accurate picture of the current level of access to grade-level content students at a grade level or school are receiving. Assessments such as classroom performance, teacher interviews, and universal screeners can provide the necessary data for a student access survey.

Where Are They Now?

Identify the percentage of current students working above, at, or below grade level.

Grade Level:	Number of Students	Percentage of Class
Working daily above grade level		
Working daily at grade level		
Working daily at grade level (and also receiving instruction below grade level)		
Working daily below grade level		

Questions to Monitor Access to Grade-Level Content

1. For what reasons are some students working below grade level?
2. For what reasons are some students working above grade level?
3. What efforts are being made this year to support students in lower levels to ensure that next year there is an increase in students working at grade level?

Figure 4.19. "Where Are They Now?" Reflection Tool

Mathematics leaders find power in these data when they look at trends over time. Longitudinal data reveal if there are increases in the number of students working at or above grade level, as well as give program leaders an awareness of how many students are actually working below grade level each year. As leaders of mathematics discuss and evaluate the impact of their instructional improvement initiatives, the data will give them an internal measure for assessing the health of their program. Figure 4.20 shows an example of how the data could be collected and analyzed over time. Time intervals could be academic years, semesters, quarters, or any other interval based on the structure of an instructional program.

Where Are They Now?
Student Access Over Time

	Interval 1		Interval 2		Interval 3		Interval 4	
	Total	%	Total	%	Total	%	Total	%
Above grade level								
At grade level								
At grade level with support								
Below grade level								

Questions to Monitor Access to Grade-Level Content Over Time

1. Over time, are the number of students gaining access to grade-level content increasing or decreasing?
2. What might be the causes of these results?
3. What system shifts or instructional improvements might increase the number of students who have access to grade-level content?

Figure 4.20. Trend Data Analysis

Effective mathematics education leaders understand, however, that access to grade-level curriculum is just the beginning. Along with access to grade-level content, students need high-quality instruction and feedback that support them as they build their mathematical understanding. Buffom et al. (2009) assert that "a school that has significantly less than 75% of its students at or above grade-level proficiency has a core program problem, not an intervention problem" (p. 74). For this reason, a school needs to monitor student performance within teacher teams or grade levels to determine the health of its core program. Use data sources such as grades, benchmark assessments, or student work samples. Achievement data analysis protocols such as *Here's What, So What, Now What* (Figure 4.8) as seen earlier in this chapter may also be applied to data analyses at the district level.

A complicating factor is the tenuous correlation between students' classroom grades and their achievement as measured on standardized tests. Research conducted over time shows that there is, at best, a moderate correlation between classroom grades and standardized test scores (Bowers, 2019). Classroom grades are an important mechanism for providing students with feedback (Marzano, 2000), but there are additional nonacademic factors influencing classroom grades, such as district policies and individual teacher grading practices. To balance these factors, effective mathematics education leaders include multiple layers of data when assessing the health of their instructional programs.

Summary of Essential Actions

Monitoring and acting on evidence of student learning is an essential part of the cycle to improve student mathematics learning. Effective mathematics education leaders intentionally collect data during the work and then purposefully analyze and reflect on the data in order to take subsequent action (NCSM, 2019a). Mathematics education leaders may monitor student learning at the classroom level by

- providing opportunities for teachers to use and design formative assessment strategies,
- providing opportunities for teachers and teacher teams to analyze data generated from formative assessments, and
- helping teachers create structures to respond to formative assessment data in subsequent classroom instruction.

Mathematics education leaders may monitor student learning in teacher teams by

- leading the creation of high-quality common classroom assessments,
- creating structures that support data analysis from the results of the assessments, and
- creating structures for student reflection and goal setting.

Mathematics education leaders may monitor student learning at the district level by

- leading the creation of high-quality common district assessments,
- monitoring the strategic implementation of student intensification plans, and
- monitoring students' access to grade-level content.

Taking Action

Figure 4.21 is a reflection on the three imperatives aligned to the fourth guiding principle: Monitor and act on evidence of student learning. Complete the reflection to identify strengths and areas of possible growth.

Action	Reflection Questions	Comments
Imperative 1: *Ensure the design and use of high-quality, aligned assessments and equitable assessment processes that guide meaningful reflection and action.*		
Effective mathematics leaders facilitate the use and development of meaningful formative assessments.	How do teachers currently use formative assessment? What do teachers need to know more about in order to use formative assessment more effectively?	
Effective mathematics leaders ensure that teachers use high-quality common assessments that promote equitable access to grade-level content.	How well do teacher teams use common assessments?	
Effective mathematics leaders design and monitor high-quality, standards-based common assessments and use their results to guide meaningful reflection and action, particularly regarding equitable intensification strategies.	How well do our district-level common assessments align to state standards? How do we use data from these common assessments to inform our instructional programs?	
Imperative 2: *Ensure that the evidence of learning collected from every assessment is used to inform the design of curriculum, instruction, and the assessments themselves.*		
Effective mathematics leaders ensure that teachers analyze data generated from classroom-based formative assessments.	What structures do we have in place that support teachers' analysis of data?	
Effective mathematics leaders ensure that teacher teams analyze data generated from high-quality common assessments and use the data to inform curriculum, instruction, and the assessments themselves.	What tools do teacher teams use to analyze data? How are teachers analyzing the data to ensure that all students have equitable access to grade-level content?	
Effective mathematics leaders ensure that intensification strategies meet students' needs by monitoring data and using that data to inform intensification plans for all students.	What data do we have available to create intensification plans? How do we know the right students are in the right intensification strategy?	
Imperative 3: *Ensure every student is provided content based on grade-level content and intensification based on evidence of student learning.*		
Effective mathematics leaders ensure that teachers respond to formative assessment data by intervening during instruction.	What knowledge do teachers need in order to effectively change instruction based on formative assessment data?	
Effective mathematics leaders create structures for student reflection and goal setting.	What structures do we currently have in place that involve students in setting learning and achievement goals?	
Effective mathematics leaders monitor every student's access to grade-level content and use data to develop systemic changes when necessary.	How do we ensure that every student has access to grade-level content? How are students selected for particular intensification strategies? Are those selections equitable?	

Figure 4.21. Self-Reflection on Guiding Principle 4

Visit www.mathedleadership.org/EAresources to download a free reproducible version of this figure.

In this chapter, you learned more about how effective mathematics education leaders may monitor and use evidence of student learning in order to inform instructional practices, curriculum design, and district- or campus-level policies regarding students' access to high-quality instruction and any necessary instructional supports. The next and final chapter of this book presents a fuller picture of the *Framework for Leadership in Mathematics Education.*

CHAPTER 5
Framing the Fuller Picture

In previous chapters, you examined the big ideas that support the four guiding principles in greater depth. You also reflected on how each of the big ideas supports the work of the three imperatives: *commitment to self, commitment to colleagues,* and *commitment to others.*

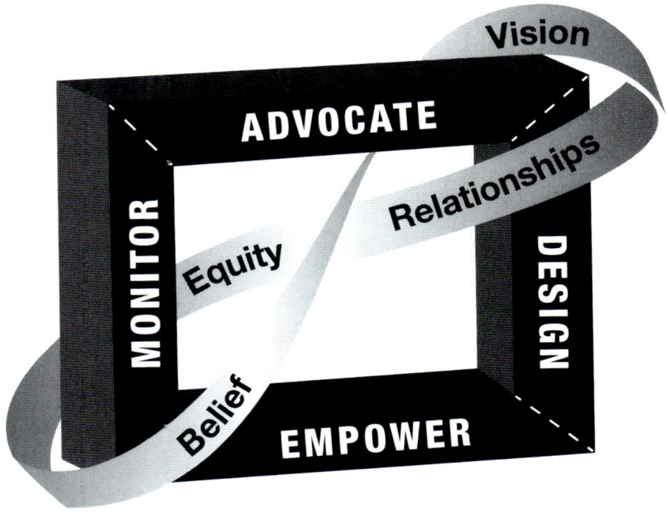

Based on what you know now, take a few minutes and use the reflective tool in Figure 5.1 to record some commitments that you can make, based on your leadership role, to take action and begin, or continue, the work toward goals that support each of the guiding principles.

Guiding Principle	Reflection
In your leadership role, how might you advocate for and expect high-quality, equitable mathematics teaching and learning for every student?	
In your leadership role, how might you design and implement structures that support high-quality mathematics teaching and learning in your classroom?	
In your leadership role, how might you empower and nurture a culture of productive professionalism?	
In your leadership role, how might you monitor and act on evidence of student learning?	

Figure 5.1. Role-Based Reflective Tool

Throughout the book, we analyzed 12 essential actions emerging from the intersection of four guiding principles and three imperatives, shown in Figure 5.2.

Guiding Principle	Imperative 1 *Commitment to Self*	Imperative 2 *Commitment to Colleagues*	Imperative 3 *Commitment to Others*
ADVOCATE and expect high-quality, equitable mathematics teaching and learning for every student.	*Ensure that every teacher possesses a shared understanding and vision of high-quality mathematics instruction and the actions required to meet the vision.*	*Ensure that every teacher possesses the skills and knowledge necessary to design and implement meaningful learning experiences that lead to student understanding of mathematics.*	*Ensure that all stakeholders have a clear understanding of high-quality mathematics instruction and how to support it.*
DESIGN and implement structures that support high-quality mathematics teaching and learning for every teacher.	*Ensure mathematics learning for all students through organizational structures, time and resource allocation, and systemic supports that are aligned, intentional, and equitable.*	*Ensure systems of continual collaborative, job-embedded professional learning to build teacher and leader capacity and increase efficacy.*	*Ensure sustainability through engaging all stakeholders in systemic, long-range strategic planning for all teaching and learning improvement initiatives.*
EMPOWER and nurture a culture of productive professionalism.	*Ensure assumptions, beliefs, expectations, and habits are examined in order to shape the school or department culture around teaching and learning of mathematics.*	*Ensure a culture of reflection, refinement, and action focused on continuous improvement in mathematical learning.*	*Ensure students, teachers, families, and community partnerships are built upon meaningful engagement.*
MONITOR and act on evidence of student learning.	*Ensure the design and use of high-quality, aligned assessments and equitable assessment processes that guide meaningful reflection and action.*	*Ensure that the evidence of learning collected from every assessment is used to inform the design of curriculum, instruction, and the assessments themselves.*	*Ensure every student is provided access to grade-level content and intensification based on evidence of student learning.*

Figure 5.2. Framework for Leadership in Mathematics Education

These essential actions are anchored in four NCSM foundational elements: *beliefs, vision, equity,* and *relationships.* A mathematics education leader has a responsibility to examine and facilitate alignment of all stakeholders' **beliefs** regarding mathematics education so that they are aligned with a shared **vision** of high-quality teaching and learning that aligns with what we know about research and better practices. All established systems and structures must include a focus on **equity** so that all teachers and students may be served by those structures. This work is only possible if we collaborate with one another and establish positive working **relationships** with all stakeholders.

When leaders **advocate** and expect high-quality, equitable mathematics teaching and learning for every student, the beliefs of all stakeholders must be examined so that the work may move forward. The advocacy and expectations must help move everyone toward the shared vision of high-quality teaching and learning using systems and structures with an eye on equity. We also know from research (DuFour & Eaker, 1998; Kanold & Larson, 2015) that the most effective work is done collaboratively in teacher teams. Thus, effective mathematics education leaders know that they must build relationships in their advocacy.

The **design** and implementation of structures that support high-quality mathematics teaching and learning for every teacher requires attention to the beliefs of all stakeholders because if teachers and leaders that you work with do not believe the structures will get results, then they will not be implemented effectively. These structures must emerge from the shared vision for the same reason. With an increasingly diverse student and teacher population, we must keep a laser-sharp focus on equity so that all students may have the opportunity to engage in meaningful and relevant mathematics. All of this work relies on positive working relationships to accomplish the task.

One of the most important actions a mathematics education leader can do to sustain the work is to **empower** and nurture a culture of productive professionalism. Leaders must project a belief that all teachers and leaders in the system are professionals and capable of contributing to the work. A culture of productive professionalism is essential for moving toward a shared vision of equitable mathematics teaching and learning. Leaders must leverage relationships in order to build a collaborative culture.

For mathematics education leaders to effectively **monitor** and act on evidence of student learning, they must have data collection systems in place that align with the shared vision of teaching and learning so that teachers and leaders know how they are moving toward that vision. Part of the shared vision includes an alignment of beliefs. As teacher teams and leaders analyze evidence of student learning, they must resist temptation to engage in deficit thinking about "those kids" and instead focus on equitable access to on-grade level content and appropriate intensification plans. Collaborative analysis and decision making require positive working relationships.

Where is a mathematics education leader to begin? Large tasks are accomplished by taking a series of smaller actions. To take that first action, use the tool in Figure 5.3 to reflect on where you and your immediate colleagues are in relation to the foundational elements.

Foundational Element and Essential Actions	Description of Each Foundational Element or Essential Action	Observable Evidence to Support Each Element or Essential Action	Possible Next Steps
Beliefs	Beliefs are the core elements of educational change. Positive beliefs include the following: • We believe all students can learn. • There is a correlation between student learning and teacher practices. • Growth mindset is essential.		
Vision	Vision communicates a common direction for a mathematics program, school, district, state, or province. Vision statements are collectively created and owned by all stakeholders.		
Equity	Equity is creating the conditions, structures, and policies necessary to ensure that all students have relevant, meaningful, and high-quality sets of mathematics experiences, facilitated through effective teaching and learning practices with high expectations and the support and resources needed to maximize student learning.		
Relationships	Coaching requires establishing, nurturing, and maintaining positive, productive, and collaborative relationships. Continually nurturing and building trust is essential for a coaching relationship.		

Figure 5.3. Reflection Tool for Foundational Elements

Visit www.mathedleadership.org/EAresources to download a free reproducible version of this figure.

The next bite is to engage in a Plan-Do-Study-Act cycle with the colleagues you lead. This cycle is frequently done in school improvement planning and gives a collaborative teacher or leadership team a tangible plan that can guide their work.

Figure 5.4 is a starting point for developing a strategic plan and applying research and resources to leadership work. During the strategic planning process, we will keep the following in mind:

1. Involve all stakeholders to include a wide variety of perspectives.
2. Reflect on how the goals align with the "vision statement."
3. Be prepared to revisit this plan each year and adjust strategic priorities and goals as needed.

<table>
<tr><th colspan="2">Process</th></tr>
<tr><td>Plan</td><td>The Plan phase is sometimes referred to as the Assess and Analyze phase. Leaders will answer the following questions: (1) Where do we/I want to go? (2) Where are we now? and (3) How do we get there?
• Teams will develop a vision statement of high-quality mathematics through the lens of coaching. This statement is intended to influence thinking and decisions.
• Create a list of strengths and weaknesses in achieving the vision and create a list of opportunities and challenges.
• Analyze current reality in terms of student achievement.
• Set goals/targets and develop strategic priorities that will help us achieve our vision statement.
• Identify annual goals that support each of the top strategic priorities (3- to 5-year plan).
• Determine the timeline, resources, strategies, and people necessary to meet the annual goals under each of the strategic priorities.
• Determine milestones to ensure everyone stays on course while implementing the plan.</td></tr>
<tr><td>Do</td><td>The Do phase is where all of the action takes place.
• Implement strategies and engage in professional learning.
• Provide ongoing feedback and time for reflection.</td></tr>
<tr><td>Study</td><td>The Study phase is sometimes referred to as the Monitor phase.
• Gather and review relevant data.
• Regularly monitor progress and recommend revisions to the plan, if needed.
• Continually evaluate all decisions to make sure they support the plan, providing feedback to the strategic planning team on their implementation.</td></tr>
<tr><td>Act</td><td>The Act phase is sometimes referred to as the Reflection or Revision phase.
• Assess progress toward goals or targets.
• Review the strategic plan annually with all members, including the vision statement, strategic priorities, and annual goals, and revise it as needed.
• Repeat the full strategic planning process every 3 to 5 years to create a new plan or uphold the current one.</td></tr>
</table>

Figure 5.4. Plan-Do-Study-Act Cycle of Reflection and Action

Visit www.mathedleadership.org/EAresources to download a free reproducible version of this figure.

Mathematics education leadership has rarely been more relevant or more required than it is today. Increasingly complex technologies generate volumes of data describing teacher and student performance. The *Framework for Leadership in Mathematics Education* will help you in your leadership role as you seek to make sense of these data and support initiatives that will benefit all students in your care.

> *NCSM is a mathematics education leadership organization that equips and empowers a diverse education community to engage in leadership that supports, sustains, and inspires high-quality mathematics teaching and learning every day for each and every learner.*

We began this book with a look at the NCSM Mission Statement. It is appropriate to come full circle and revisit that statement at the conclusion of this book. We believe that the 12 essential actions emerging from the four guiding principles will help you in your pursuit of this mission.

We believe that these essential actions will equip and empower you, a member of a diverse education community, to engage in leadership in your particular context and in your particular role, which supports, sustains, and inspires high-quality mathematics teaching and learning every day that each and every learner in your school, system, district, province, or state deserves.

NCSM Past President John Staley (2015–2017) challenged NCSM members at the NCSM 2016 Annual Meeting to consider their "why." In that spirit, after you have studied the *Framework for Leadership in Mathematics Education* and are preparing to act on that framework, envision a child who is significant in your life. That child could be your own child, a niece or nephew, or a neighbor. Think about that child's present and your hopes for that child's future. As you take action, ask yourself how that action will create a better future for that child. The decisions you make and the actions that you take will impact a child who is significant in someone's life. We should demand the same high-quality teaching and learning for each and every child that families in our communities have entrusted to our care.

We believe that the *Framework for Leadership in Mathematics Education* will help you in that pursuit.

References

Aguilar, E. (2013). *The art of coaching: Effective strategies for school transformation.* San Francisco, CA: Jossey-Bass.

Aguirre, J., Martin, D., & Mayfield-Ingram, K. (2013). *The impact of identity in K–8 mathematics learning and teaching: Rethinking equity-based practices.* Reston, VA: NCTM.

Aguirre, J. M., & Zavala, M. (2013). Making culturally responsive mathematics teaching explicit: A lesson analysis tool. *Pedagogies: An International Journal, 8*(2), 163–190.

Annie E. Casey Foundation. (2017). *Race for results—Building a path to opportunity for all children* [Policy report]. Retrieved from https://www.aecf.org/m/resourcedoc/aecf-2017raceforresults-2017.pdf

Baldinger, E. M., Louie, N., & the Algebra Teaching Study and Mathematics Assessment Project. (2016). *The TRU Math conversation guide: A tool for teacher learning and growth* Berkeley, CA: Graduate School of Education, University of California, Berkeley & College of Education, Michigan State University. Retrieved from https://truframework.org/

Ball, D. L., & Cohen, D. K. (1996). Reform by the book: What is—or might be—the role of curriculum materials in teacher learning and instructional reform? *Educational Researcher, 25*(9), 6–8, 14.

Ball, D. L., Hill, H. C., & Bass, H. (2005). Knowing mathematics for teaching: Who knows mathematics well enough to teach third grade, and how can we decide? *American Educator, 30*(3), 14–17, 20–22, 43–46.

Bandura, A. (1996). *Self-efficacy in changing societies.* New York, NY: Cambridge University Press.

Barnes, B., & Toncheff, M. (2016). *Activating the vision: The four keys of mathematics leadership.* Bloomington, IN: Solution Tree Press.

Bishop, R. S. (1990, Summer). Mirrors, windows, and sliding glass doors. *Perspectives: Choosing and Using Books for the Classroom, 6*(3). Retrieved from https://scenicregional.org/wp-content/uploads/2017/08/Mirrors-Windows-and-Sliding-Glass-Doors.pdf

Boaler, J. (2015). *Mathematical mindsets: Unleashing students' potential through creative math, inspiring messages and innovative teaching.* New York, NY: Jossey-Bass.

Bowers, A. J. (2019). *Report card grades and educational outcomes.* In T. R. Guskey & S. M. Brookhart (Eds.), *What we know about grading: What works, what doesn't, and what's next.* Alexandria, VA: ASCD.

BrainyQuote.com. (n.d.). Audre Lorde quotes. Retrieved November 14, 2019, from https://www.brainyquote.com/quotes/audre_lorde_357287

Briceño, S. (2018, January 30). *8 Ways to help your students develop a growth mindset in math* [Blog entry]. Retrieved from https://www.carnegielearning.com/blog/8-ways-to-help-students-develop-growth-mindset-in-math/

Brown, B. (2018). *Dare to lead: Brave work. Tough conversations. Whole hearts.* New York, NY: Random House.

Buffom, A., Mattos, M., & Weber, C. (2009). *Pyramid response to intervention: RTI, professional learning communities, and how to respond when kids don't learn.* Bloomington, IN: Solution Tree Press.

Carr, J. F., Herman, N., & Harris, D. E. (2005). *Creating dynamic schools through mentoring, coaching, and collaboration.* Alexandria, VA: ASCD.

CCSSO & NGA. (2010). *Common Core State Standards for Mathematics.* Retrieved from http://www.corestandards.org/faq/what-are-educational-standards/

Charles A. Dana Center. (2019). *The case for mathematics pathways* [White paper]. Retrieved from https://dcmathpathways.org/sites/default/files/resources/2019-03/CaseforMathPathways_20190313.pdf

Cheung, R., Reinhardt, T., Stone, E., & Little, J. W. (2018). Defining teacher leadership: A framework. *Phi Delta Kappan, 100*(3), 38–44.

Confrey, J. (2010). "Both and"—equity and mathematics: A response to Martin, Gholson, and Leonard. *Journal of Urban Mathematics Education, 2*(3), 25–33.

Delaware Math Coalition. (2013). *Old normal: New normal.* Retrieved from http://delawaremath.dreamhosters.com/new-normal/

DesMarais, C. (2017, September 9). *Science says the most successful kids have parents who do these 9 things* [Blog post]. Retrieved from https://www.inc.com/christina-desmarais/want-to-raise-kids-who-thrive-science-says-do-thes.html

DuFour, R. (2016). *In praise of American educators.* Bloomington, IN: Solution Tree Press.

DuFour, R., & Eaker, R. (1998). *Professional learning communities at work: Best practices for enhancing student achievement.* Bloomington, IN: National Educational Service.

DuFour, R., & Marzano, R. (2011). *Leaders of learning: How district, school, and classroom leaders improve student achievement.* Bloomington, IN: Solution Tree Press.

Dweck, C. S. (2016). *Mindset: The new psychology of success* (2nd ed.). New York, NY: Random House.

Emdin, C. (2016). *For white folks who teach in the hood...and the rest of y'all too: Reality pedagogy and urban education.* Boston, MA: Beacon Press.

Feldman, J. (2019). Beyond standards-based grading: Why equity must be part of grading reform. *The Kappan, 100*(8), 52–55.

Fennell, F. (2007). For principals ONLY! NCTM President Message. *NCTM News Bulletin.* Retrieved from https://www.nctm.org/News-and-Calendar/Messages-from-the-President/Archive/Skip-Fennell/For-Principals-ONLY!/

Flaum, J. P. (2010). *When it comes to business leadership, nice guys finish first.* Green Peak Partners. Retrieved from https://greenpeakpartners.com/wp-content/uploads/2018/09/Green-Peak_Cornell-University-Study_What-predicts-success.pdf

Gay, G. (2000). *Culturally responsive teaching: Theory, research, & practice.* New York, NY: Teachers College Press.

Gay, G., & Kirkland, K. (2003). Developing cultural critical consciousness and self-reflection in preservice teacher education. *Theory Into Practice, 42*(3), 181–187.

Google. (n.d.). Bold definition. Retrieved June 25, 2019, from https://www.google.com/search?q=Dictionary

Gover, E. (2017, January 23). *How social media enhances teacher professional development: Interviews with experts* [Blog post]. Retrieved from https://rossieronline.usc.edu/blog/teacher-interviews-social-media/

Gutiérrez, R. (2002). Enabling the practice of mathematics teachers in context: Toward a new equity research agenda. *Mathematical Thinking and Learning, 4*(2–3), 145–187.

Gutiérrez, R. (2012). Embracing Nepantla: Rethinking "knowledge" and its use in mathematics teaching. *REDIMAT—Journal of Research in Mathematics Education, 1*(1), 29–56.

Gutiérrez, R. (2013). The sociopolitical turn in mathematics education. *Journal for Research in Mathematics Education, 44*(1), 37–68.

Harbin Miles, R., Kobett, B. M., & Williams, L. A. (2019). *The mathematics lesson-planning handbook, Grades 3–5: Your blueprint for building cohesive lessons.* Thousand Oaks, CA: Corwin.

Hattie J., Fisher, D., & Frey, N. (2017). *Visible learning for mathematics: What works best to optimize student learning.* Thousand Oaks, CA: Corwin.

Jilk, L. M. (2016). Supporting teacher noticing of students' mathematical strengths. *Mathematics Teacher Educator, 2*(4), 188–199.

Jones, S.M. (2018). Teachers' conceptions of teaching mathematics for social justice: Building on their knowledge of culturally relevant pedagogy, *New England Mathematics Journal, 51*(2), 50-63.

Joyce, B., & Showers, B. (2002). *Student achievement through staff development* (3rd ed.). Alexandria, VA: Association for Supervision and Curriculum Development.

Kanold, T. D. (2017). *HEART! Fully forming your professional life as a teacher and leader.* Bloomington, IN: Solution Tree Press.

Kanold, T. D., & Larson, M. R. (2012). *Common core mathematics in a PLC at work: Leader's guide.* Bloomington, IN: Solution Tree Press.

Kanold, T. D., & Larson, M. R. (2015). *Beyond the common core: A handbook for mathematics in a PLC at Work™. Leaders' guide.* Bloomington, IN: Solution Tree Press.

Kanold, T. D., Schuhl, S., Barnes, B., Larson, M., Kanold-McIntyre, J., & Toncheff, M. (2018). *Mathematics assessment & intervention in a PLC at work.* Bloomington, IN: Solution Tree Press.

Knight, J. (2011). *Unmistakable impact: A partnership approach for dramatically improving instruction.* Thousand Oaks, CA: Corwin.

Kreisberg, H., & Beyranevand, M. L., (2019). *Adding parents to the equation: Understanding your child's elementary school math.* Lanham, MD: Rowman & Littlefield.

Ladson-Billings, G. (1994). *The dreamkeepers: Successful teachers of African-American children.* San Francisco, CA: Jossey-Bass.

Lambert, L. (1995). Toward a theory of constructivist leadership. In L. Lambert (Ed.), *The constructivist leader.* New York, NY: Teachers College Press.

Lambert, L. (2002). A framework for shared leadership. *Educational Leadership, 59*(8), 37–40.

Larson, M. R., & Kanold, T. D. (2016). *Balancing the equation: A guide to school mathematics for educators and parents.* Reston, VA: NCTM.

Leinwand, S. (2012). *Sensible mathematics: A guide for school leaders in the era of Common Core State Standards* (2nd ed.). Portsmouth, NH: Heinemann.

Lieberman, A., & Miller, L. (2004). *Teacher leadership.* San Francisco, CA: Jossey-Bass.

Loucks-Horsley, S., Stiles, K. E., Mundry, S., Love, N., & Hewson, P. W. (2010). *Designing professional development for teachers of science and mathematics* (3rd ed.). Thousand Oaks, CA: Corwin.

Lynch, K., Hill, H. C., Gonzalez, K., & Pollard, C. (2019). *Strengthening the research base that informs STEM instructional improvement efforts: A meta-analysis.* Working paper, Brown University.

Lynn, J., & Briars, D. (2011, October). *Reclaiming lost ground: Research-based interventions for underprepared algebra students.* Paper presented at the 2011 NCTM Regional Conference, Atlantic City, NJ.

Macmillan. (n.d.). Bold definition. Retrieved June 25, 2019, from https://www.macmillandictionary.com/us/dictionary/american/bold_1

Marzano, R. J. (2000). *Transforming classroom grading.* Alexandria, VA: ASCD.

Marzano, R. J. (2003). *What works in schools: Translating research into action.* Alexandria, VA: ASCD.

Marzano, R. J. (2006). *Classroom assessment & grading that work.* Alexandria, VA: ASCD.

Matthews, L. E., Jones, S. M., & Parker, Y. A. (2013). Advancing a framework for culturally relevant, cognitively demanding mathematics tasks. In J. Leonard & D. Martin (Eds.), *The brilliance of Black children in mathematics: Beyond the numbers and toward a new discourse.* Charlotte, NC: Information Age Publishing.

Maxwell, J. C. (2011). *The 5 levels of leadership.* New York, NY: Center Street.

McGatha, M. B., Bay-Williams, J. M., Kobett, B. M., & Wray, J. A. (2018). *Everything you need for mathematics coaching: Tools, plans, and a process that works for any instructional leader.* Thousand Oaks, CA: Corwin.

Meir, G. (n.d.). *Wise old sayings.* Retrieved from http://www.wiseoldsayings.com/be-bold-quotes/

Merriman, M. (2015). *What if the next big disruptor isn't a what but a who?* Ernst & Young. Retrieved April 28, 2017, from http://www.ey.com/Publication/vwLUAssets/ EY-rise-of-gen-znew-challenge-forretailers/$FILE/EY-rise-of-gen-znewchallenge-for-retailers.pdf

Moll, L. C., Amanti, C., Neff, D., & Gonzalez, N. (1992, Spring). Funds of knowledge for teaching: Using a qualitative approach to connect homes and classrooms. *Theory Into Practice, 31*(2), 132–141.

National Council of Teachers of Mathematics (NCTM). (2000). *Principles and standards for school mathematics.* Reston, VA: Author.

National Council of Teachers of Mathematics (NCTM). (2006). *Curriculum focal points for prekindergarten through Grade 8 mathematics: A quest for coherence.* Reston, VA: Author.

National Council of Teachers of Mathematics (NCTM). (2014a). *Principles to actions: Ensuring mathematical success for all.* Reston, VA: Author.

National Council of Teachers of Mathematics (NCTM). (2014b). *Access and equity in mathematics education* [Position statement]. Retrieved from https://www.nctm.org/uploadedFiles/Standards_and_Positions/Position_Statements/Access_and_Equity.pdf

National Council of Teachers of Mathematics (NCTM). (2017). T*aking action: Implementing effective teaching practices.* Reston, VA: Author.

National Council of Teachers of Mathematics (NCTM). (2018). *Catalyzing change for high school mathematics: Initiating critical conversations.* Reston, VA: Author.

National Council of Teachers of Mathematics (NCTM). (2019). *Work to end tracking and offer four years of meaningful math instruction* [Case study]. Retrieved from https://www.nctm.org/uploadedFiles/Standards_and_Positions/Work-to-End-Tracking-and-Offer-Four-Years-of-Meaningful-Final.pdf

National Research Council. (2001). *Adding it up: Helping children learn mathematics.* Washington, DC: National Academies Press.

NCSM. (2007). *Improving student achievement by leading effective and collaborative teams of mathematics teachers* [Position paper]. Retrieved from https://www.mathedleadership.org/resources/position.html

NCSM. (2008). *The PRIME leadership framework: Principles and indicators for mathematics education leaders.* Bloomington, IN: Solution Tree Press.

NCSM. (2009). *Improving student achievement in mathematics by leading highly effective assessment practices.* [Position paper]. Retrieved from https://www.mathedleadership.org/resources/position.html

NCSM. (2010). *Improving student achievement in mathematics by promoting positive self-beliefs* [Position paper]. Retrieved from https://www.mathedleadership.org/resources/position.html

NCSM. (2013). *Improving student achievement by infusing highly effective instructional strategies into RtI tier 1 instruction* [Position paper]. Retrieved from https://www.mathedleadership.org/resources/position.html

NCSM. (2014a). *Improving student achievement in mathematics through formative assessment in instruction* [Position paper]. Retrieved from https://www.mathedleadership.org/resources/position.html

NCSM. (2014b). *It's TIME: Themes and imperatives for mathematics education* Bloomington, IN: Solution Tree Press.

NCSM. (2015). *Improving student achievement by infusing highly effective instructional strategies into multi-tiered support systems (MTSS)—Response to intervention (RtI) Tier 2 instruction* [Position paper]. Retrieved from https://www.mathedleadership.org/resources/position.html

NCSM. (2019a). *NCSM essential actions: Coaching in mathematics education.* Englewood, CO: Author.

NCSM. (2019b). *NCSM essential actions: Instructional leadership in mathematics education.* Englewood, CO: Author.

NCSM. (2019c). *Acknowledging our role in the education debt: Working to transform school mathematics for learners of color.* Englewood, CO: Author.

NCSM. (2019d). *Framework for leadership in mathematics education: Executive summary.* Englewood, CO: Author.

NCSM. (in press). *Closing the opportunity gap with mathematics course pathways and policies: The need for detracking.* Retrieved from http://www.mathedleadership.org

NCSM & TODOS. (2016). *Mathematics education through the lens of social justice: Acknowledgment, actions, and accountability.* Retrieved from https://www.mathedleadership.org/docs/resources/positionpapers/NCSMPositionPaper16.pdf

Noguera, P. A. (2018, March). *Equity and deeper learning in mathematics.* Keynote presentation at the NCSM Annual Conference, Washington, DC.

O'Brien, D. (2019). *Teaching parents a growth mindset* [Blog post]. Retrieved from https://dynamath.scholastic.com/pages/dynamath-expressions/2018-19/teaching-parents-a-growth-mindset.html

O'Leary, S. (2017). *Parenting in the real world: The rules have changed. Drop the guilt. Handle any parenting situation in 7 simple steps.* Author published.

Paris, D., & Alim, H. S. (2017). *Culturally sustaining pedagogies: Teaching and learning for justice in a changing world.* New York, NY: Teachers College Press.

Reeves, D. B. (2004). *Assessing educational leaders: Evaluating performance for improved individual and organizational results.* Thousand Oaks, CA: Corwin.

Reeves, D. B. (2009). *Leading change in your school: How to conquer myths, build commitment, and get results.* Alexandria, VA: ASCD.

Reeves, D. B. (2010). *Transforming professional development into student results.* Alexandria, VA: ASCD.

Rich, A. (1986). *Blood, bread, and poetry: Selected prose 1979–1985.* New York, NY: W. W. Norton.

Seeley, C. L. (2009). *Faster isn't smarter: Messages about math, teaching, and learning in the 21st century.* Sausalito, CA: Math Solutions.

Sergiovanni, T. J. (1992). *Moral leadership: Getting to the heart of school improvement.* San Francisco, CA: Jossey-Bass.

Sheldon, S. B. (2011). A framework for partnerships. In S. Redding, M. Murphy, & P. Sheley (Eds.), *Handbook on family and community engagement.* Lincoln, IL: Academic Development Institute.

Shulman, L. S. (1986). Those who understand: Knowledge growth in teaching. *Educational Researcher, 15*(2), 4–14.

Smith, M. S., & Stein, M. K. (1998). Selecting and creating mathematical tasks: From research to practice. *Mathematics Teaching in the Middle School, 3*(5), 344–349.

Sparks, S. D. (2019, March 12). Why teacher-student relationships matter: New findings shed light on best approaches. *Education Week, 38*(25), 8.

Stiggins, R. (2007). "Assessment for learning: An essential function of productive instruction. In D. Reeves (Ed.), *Ahead of the curve: The power of assessment to transform teaching and learning.* Bloomington, IN: Solution Tree Press.

Stiggins, R. (2014). *Revolutionize assessment.* Thousand Oaks, CA: Corwin.

Stiggins, R. J., Arter, J., Chappuis, J., & Chappuis, S. (2006). *Classroom assessment for student learning: Doing it right—using it well.* Princeton, NJ: Educational Testing Service.

Toncheff, M. (2018, June 18). *What is your mathematics team's language about student learning?* [Blog post]. Retrieved from https://www.solutiontree.com/blog/changing-negative-math-team-talk/

Turner, A. (2018). *The business case for racial equity—A strategy for growth.* W. K. Kellogg Foundation [Report]. Retrieved from https://www.wkkf.org/resource-directory/resource/2018/07/business-case-for-racial-equity

Vagle, N. M. (2009). *Finding the meaning in numbers.* In T. R. Guskey (Ed.), *The principal as assessment leader.* Bloomington, IN: Solution Tree Press.

Wellman, B., & Lipton, L. (2004). *Data-driven dialogue: A facilitator's guide to collaborative inquiry.* Sherman, CT: MiraVia, LLC.

Wiggins, G., & McTighe, J. (2005). *Understanding by design* (2nd ed.) Alexandria, VA: ASCD.

Wiliam, D. (2016). *Leadership for teacher learning: Creating a culture where all teachers improve so that all students succeed.* West Palm Beach, FL: Learning Sciences International.